Wilderness, Wildlands, and People

A Partnership for the Planet

Edited by

Vance G. Martin and Cyril F. Kormos

8th WORLD WILDERNESS CONGRESS
PROCEEDINGS

Fulcrum Publishing
Golden, Colorado

The WILD Foundation
Boulder, Colorado

Library of Congress Cataloging-in-Publication Data

World Wilderness Congress (8th : 2005 : Anchorage, Alaska)
  Wilderness, wildlands, and people : a partnership for the planet / editor, Vance G. Martin.
      p. cm.
  Proceedings of the 8th World Wilderness Congress, held in Anchorage, Alaska, September-October 2005.
  ISBN 978-1-55591-602-2 (alk. paper)
  1. Wilderness areas--Congresses. 2. Nature conservation--Congresses. I. Martin, Vance. II. Title.
  QH75.A1W67 2008
  333.78'216--dc22

                               2007039785

Printed in the United States of America by Color House Graphics, Inc.
0 9 8 7 6 5 4 3 2 1

Design by Patty Maher
Interior watercolor illustrations courtesy of Rod McIver and *Heron Dance*
World maps by Darin Sanders

Fulcrum Publishing
4690 Table Mountain Drive, Suite 100
Golden, Colorado 80403
800-992-2908 • 303-277-1623
www.fulcrumbooks.com

Contents

vened in 1977 in South Africa, and has since met on seven more occasions around the globe and is now the world's longest-running, international environmental forum.

The WWC does much more than bring diverse, concerned people together to share information. An ongoing conservation initiative with distinct three- to four-year project cycles, the WWC sets practical objectives and creates a process to achieve them. When each Congress actually convenes, delegates present, assess, debate, elaborate upon, endorse, and celebrate these conservation results. The WILD Foundation is the originator, organizer, and convenor of this process, but the impressive and targeted list of WWC accomplishments is the work of very many dedicated people, sponsors, organizations, and institutions.

The 8th WWC built on that proud tradition of creating identifiable outcomes. Scientists, Native people, politicians, corporate leaders, artists, educators, managers, and others reviewed the first wilderness area in Latin America made possible by Mexico's pioneering wilderness law, also the first such law in Latin America. Delegates attended many training programs; expanded the list of private sector wilderness areas; convened the first Native Lands and Wilderness Council; created the International League of Conservation Photographers; critiqued new wilderness inventories and maps; and much more. Details are within these pages, with a summary of the practical accomplishments in the appendices.

The achievements of the 8th WWC are one measure of the respect that human society must have for its wild origins and the understanding we need of the critical role that wilderness plays in a future that promises health, prosperity, and sanity. But equally important to these practical accomplishments was the palpable, positive spirit of cooperation and commitment that pulsed through every conversation, debate, meeting, and announcement. Given the enormity of the challenges we face today, what can be more important than a strong sense of spirit and shared endeavor?

Despite predictions of a foreboding future, we actually live in historic times and have many more opportunities than challenges. To realize these opportunities, the WWC process and practice can perhaps be considered a good prescription for human society as a whole.

First, understand how wilderness builds its strength and makes possible all life on Earth, providing irreplaceable services generated through component linkages, system interrelationships, and recurring yet different cycles.

Second, respect diversity, listen deeply to each other, and think out-of-the-box.

Third, implement the inspirations, ideas, lessons, and actions that arise from this process in a step-by-step daily process of change.

A new world is entirely within our power to create and our destiny to experience.

Boulder, Colorado, USA
March 2007

CHAPTER 1

The Wilderness Perspective—
Opening

"Wilderness"

~

IAN MCCALLUM

"Ecological Intelligence," *Africa Geographic*, 2005

Have we forgotten
that wilderness is not a place,
but a pattern of soul
where every tree, every bird and beast
is a soul maker?

Have we forgotten
that wilderness is not a place
but a moving feast of starts,
footprints, scales and beginnings?

Since when
did we become afraid of the night
and that only the bright stars count?
or that our moon is not a moon
unless it is full?

By whose command
were the animals
through groping fingers,
one for each hand,
reduced to the big and little five?

Have we forgotten
that every creature is within us
carried by tides
of Earthly blood
and that we named them?

Have we forgotten
that wilderness is not a place,
but a season
and that we are in its
final hour?

Wilderness and the Human Soul

Ian Player
Founder, World Wilderness Congress

A recent flight from South Africa took me over the Drakensberg Mountains, *Ukuhlamba* to the Zulu people. I looked down and pondered; saw the red grass glowing luminously in the afternoon sun. These mountains were the last refuge of the San people, or Bushmen, who painted their exquisite art on cave walls and recorded the history of our country, the coming of the Nguni people, the Boers on their horses and English soldiers, and the vast array of wild animals. By 1870 there were no San people left. They were shot and killed without mercy, and with them went vast tomes of wisdom and knowledge.

A man named Richard Nelson said, "The abandonment of ethically and spiritually based relationship with nature by our western ancestors was one of the greatest and perilous transformations of the western mind." Today nearly all of modern man's ills spring from this abandonment and this is why wilderness has become so important, because it reconnects us to that ancient world.

We South Africans can be proud that our country was the first in Africa to proclaim a game reserve and the first wilderness area. Imfolozi Game Reserve in KwaZulu-Natal has that double distinction.

The World Wilderness Congress has come a long way on torturous paths, having to overcome what at times seemed insuperable odds. It has now become a critically important forum that provides a platform for many divergent views. It is important, I believe, that we look at the history of the World Wilderness Congress. Vance Martin,

President of The WILD Foundation, tells me it has now become the longest-running, public, international environmental forum. This Congress was born in South Africa in 1976, in the small wilderness area of Imfolozi Game Reserve in KwaZulu. It was a suggestion of my great friend and mentor Magqubu Ntombela, who had lead many treks into the wilderness with me. He said that we needed a big *indaba* to bring together everyone who had trekked so that we could share experiences. He was a man who could neither read nor write, but he was the wisest most gracious and bravest man I have ever known. The African people have a word for it: *ubuntu.*

It is fitting too that the World Wilderness Congress began in Africa. It is the cradle of mankind. All of us here have our origin from that mighty continent, as DNA has proved. C. G. Jung said, "We do not come into the world tabula rasa." Three million years of Africa is imprinted on the human psyche. I know from taking many hundreds of people in small groups from all over the world on foot treks into the wilderness of Imfolozi and Lake St. Lucia how they are gripped by the spirit of Africa, how at night they sleep on the red earth, dream their dreams, and hear the animals and birds. There is a connection that is evoked from the depths of the collective unconscious: the rasping cough of the leopard, the howl of the hyena, and the scream of the elephant. It is an experience that has awakened thousands of people to the value of the African wilderness and the understanding that this was once their home, which this inspires them to protect it. As Shakespeare says in Othello, "It is the cause, it is the cause my soul." And so it has become for many of us, worldwide.

In 1977, South Africa was a pariah nation, and organizing that first Congress in Johannesburg in October of that year was a nightmare. But the Congress was an undeniable success, where for the first time a black field ranger, Magqubu Ntombela, took his rightful place amongst leading international scientists, politicians, writers, and artists Bushmen (Kalahari).

It established the importance of wilderness in breaking down racial barriers in South Africa, and the wilderness trails in the Imfolozi

Game Reserve were a leading example. Ntombela used to tell the mixed groups as we sat around the fire at night, "If we are charged by rhino or lion and blood flows, it will be the same color blood for everyone, even though our skins may be a different color."

The congresses that followed in Australia, the United Kingdom, the United States, and Norway were also beset with political problems because the Congress had originated in South Africa, and because I am a South African. I will always be grateful to those American and international conservationists who stood by us and ensured that the congresses became a forum for everything associated with wilderness. Vance Martin knows this because he was at the coal face from 1983.

Today, thanks to Nelson Mandela and the peaceful elections in 1994, South Africa is the brightest light on the continent of Africa and stands poised to be a wilderness and conservation example for all of emerging Africa. But we in the world wilderness movement are under no illusions about the difficulties that lie ahead. The struggle for political freedom is over in South Africa, but not in all the African states. The new struggle is an environmental one for all our people to make wise use of the natural resources.

In 2001, the World Wilderness Congress returned to South Africa—to a transformed country. Thanks to Adrian Gardiner, Andrew Muir, and the Eastern Cape government, it was a phenomenal success. South Africa has proved what can be done.

The same is not true for other parts of Africa. I do not want to enlarge on a litany of woes facing conservation in Africa, but the problems range from the desperate situation of the last remaining Northern White Rhino in the Democratic Republic of the Congo to some parks where the game scouts do not have boots.

At the recent G8 Summit in Britain there was a focus on Africa. One can only hope that the environment will receive proper attention, because in previous aid to Africa it did not. The G8 now has a chance to rectify it.

Whereas it is correct that the birth of the World Wilderness Congress was in Africa, the honor for the establishment of national parks and wilderness areas belongs to the United States of America. It was Americans who articulated the wilderness concept and set aside wilderness areas against what at times seemed overwhelming odds. But the spirit of one of the greatest American presidents, Theodore Roosevelt, was always with them. Not for nothing did he say, "The greatest sport the world affords is aggressive fighting for the right." Yet we must remember that Frederick Courtney Selous, the great Nimrod, was his guide in Uganda. The African wilderness made a deep impression on Theodore Roosevelt, and it affected his thinking.

In my library is a book with the prosaic title of *S.1176 Hearings Before the Committee on Interior and Insular Affairs of the United States Senate*. The pages are worn thin and underlined everywhere. The cover is tattered from constant use. It has been in my possession since 1958. A most treasured book sent to me by Howard Zahniser, the then secretary of the Wilderness Society. In it I have written, "This has been the bible of the wilderness movement in South Africa." The Americans showed us the way. It is a phenomenal story of the past, the present, and the future.

One of the witnesses quoted in *S.1176* was Sigurd Olsen. He said:

> "In days to come, the wilderness concept must be clear and shining enough to capture imaginations. It must take its place as a cultural force with all expressions of man's deepest yearnings and his noblest achievements in the realm of the mind. It must be powerful enough to withstand everywhere in the world, the coming and enormous pressures of industry and population."

Talk about intimations of the future: this is it.

S.1176 is the gripping story of the blood-and-guts fight for the conservation soul of America. You realize, too, that what it is expressing is the depth of the impact that the Native Americans made on the psyche of Anglo-America. Constantly there are echoes, and one senses their spirit in the extraordinarily eloquent pleas from some of the most eminent Americans of their day.

I first came to America in 1964 as a guest of Metro Goldwyn Meyer and through Ira Gabrielson. I met Stuart Udall, Secretary of the Interior and a man proud of his Native American blood; he became a speaker at the first World Wilderness Congress. Ten minutes in his company gave me a deep and emotionally moving insight into the soul of American conservation. He reiterated that America had to be an example for the world.

The men and women who testified for wilderness in *S.1176* were heroic people, many times going against the grain and knowing that they were up against it. They warned against roads, lodges, hotels, and restaurants in the national parks. They knew they were setting an example for the world and it had to be the right one. They were unafraid to talk, determined that the Wilderness Leadership School I initiated in 1957 would emphasize that the wilderness experience was a spiritual journey. Another witness, Edwin Way Teale, said that wilderness areas are "storehouses of wildness, and wildness will become an ever-increasing spiritual need in the crowded tomorrow."

We are now in the crowded tomorrow, with a vengeance. Try a Los Angeles Freeway on what they call a quiet day.

I love America. It has always been good and inspirational for me. But I have to tell you that an article in *The New York Times* on August 29, 2005 has caused me much stress. It is entitled "Destroying the National Parks." It refers to a document that calls for the rewriting of national park rules by one of the assistant secretaries, which has been met with profound dismay in professional national park circles. This must be stopped.

Many millions of people regard national parks, forestry, and wilderness areas as sacrosanct, what difference is government to nature and human desires fit in accordingly. The United States started the national park movement and became a leader in ethics, policy, and action. It must remain so.

The wilderness work America articulated and the rest of the world has followed is practical, political, philosophical, psychological, and scientific, but at the deepest levels there are still too few people who understand it is the work of the soul. The lines of the psalm say it best: "Be still and know that I am God." And it is in the wilderness that the stillness can be found.

We have to face the fact that rampant materialism is creating havoc in our world, and wilderness areas are under threat everywhere. This has not been helped by Judaeo-Christianity; Edward Whitmont puts it succinctly, "For several centuries traditional theology has tended to create an absolute gulf between man and nature." Yet the world seems to continue as though there were no tomorrow. We have forgotten those wonderful images in the gospels that describe John the Baptist coming out of the wilderness "clothed with camel's hair with a leather belt around his waist, and he ate locusts and wild honey."

For too long there has been a cataclysmic clash between western and indigenous cultures, with the latter being the bigger loser. Sense of place and spirit of place have been destroyed.

There is terrible potential destruction to birds, landscapes, and silence in the Highlands of Scotland and other wild country in Britain with the proposed wind farms. The Wilderness Foundation United Kingdom is vigorously fighting this danger. As C. G. Jung said, "We have lost a world that once breathed with our breath and pulsed with our blood. Did the wind use to cry and the hills shout forth praise?" A cry of helplessness from indigenous people as a once-known world is swept away.

Marie-Louise von Franz, a great depth psychologist, said:

"Western civilization is in danger of building a wall of rationality in its society, which feeling cannot penetrate. Everything has to be rational and emotion is frowned upon."

This makes the poets critically important to our cause. Wilfred Owen, a First World War poet, said that all a poet can do is to warn, and that is why true poets must be truthful. Poets warn us and they inspire us. Think of W. H. Auden's words as a reflection of ecological doomsday:

"The stars are not wanted now, put out every one,
Pack up the moon and dismantle the sun. Pour away the ocean
and sweep up the wood. For nothing now can ever come to any
good."

Compare this to the inspiration of Herman Hesse:

"Sometimes, when a bird cries out,
Or when the wind sweeps through a tree
Or a dog howls in a far off farm
I hold still and listen a long time.
My soul turns and goes back to the place
Where, a thousand forgotten years ago,
The bird and the blowing wind
Were like me, and were my brothers."

Fraser Darling, the great Scottish biologist, said:

"To deprive the world of physical wilderness, would be to inflict
a grievous wound on our own kind."

My great friend the late John Aspinall, the most famous gambler
in Britain who became a conservationist and who—even when devastated by cancer of the jaw—continued to campaign and pour millions
into the saving of the gorilla and other conservation causes, said:

"I believe that wilderness is the earth's greatest treasure. Wilderness
is the bank on which all cheques are drawn. I believe our debt to
nature is total. I believe that unless we recognise this debt and renegotiate it—we write our own epitaph. I believe that there is an
outside chance to save the earth—and most of its tenants. This
outside chance must be grasped with gamblers' hands. I believe
that terrible risks must be taken and terrible passions roused before
these ends can be accomplished."

We are all engaged in a momentous struggle, and we owe it to the early pioneers to honor their visions and their achievements.

This is our task in the twenty-first century. We need something that will stir our psychic depths and touch the images of the soul. It has to surpass creeds and instantly be recognized. We must learn a new language to convey the feelings of beauty, hope, inspiration, and sacredness for humanity and all other life. We need to remember the first principle of ecology: "Everything is connected to everything else." And the wilderness experience is the spiritual spark that ignites the understanding.

Wild Time

Jay Griffiths
Author

Clocks: caging wild time. The watch, the manacle on the wrist. The calendar, imprisoning the days in square cells. Deadlines like a barbed-wire fence. Time is a wild commons but it has been enclosed and exploited for profit, just as common wildland has been privatized and enclosed.

When I wrote my book on time, *A Sideways Look at Time*, I wanted to write about what I call "wild time" and my invitation to raft down the Taku River, which runs through British Columbia and Alaska. I wanted to draw parallels between wildland and wild time. Specifically, I wanted to set the chapter in a river, because all across the world rivers and oceans are considered symbols of time itself.

Micmac society, which I'm told has no word for time, uses the river as an image for the flow of happenings. The ancient Greeks identified time with Oceanos, the divine river. "Time glides by like a stream," wrote Ovid. Time and tide rhyme in the mind of humanity (noontide, eventide) and the word "current" refers to both time and tide. In Taoist thought, the ocean is equated with the Tao, the inexhaustible source of life. So Otis Redding picked the right place, "Sittin' on the dock of the bay, wastin' ti'ai'ai'ime," for the sea is the creator of endless hours of time.

The loveliest definition of wilderness is surely "self-willed land," and as a parallel with that, time, wild time, unclocked time is self-willed. The bear on the mooch, snuffling huckleberries, snuffles in its self-willed time; it does not snuffle to schedule. The thunder has its own hours and would spit in the face of punctuality. The river's self-willed flow, its wild time, mocks the tamed suburban character of clock time.

Anyone who has been in wildlands may know a feeling of acute distaste for wearing a watch in such places and may well know the sheer glee of shucking of your watch and plunging into the river of wild time. That response of ours is a deep reaction of the human spirit: to refuse the orders of the clock, to disobey the command of the hours. For we, just as much as the bear or the river, have our self-willed hours, and a freedom of time and a sense of its wildness is one of the most fundamental, though most metaphysical, of all our freedoms.

But when those in power have wanted to deny people freedom, to control them, they have used time as part of their language of power. When the ancient Chinese empire colonized new territory, the phrase they used for this was sinister and telling: that new territory had "received the calendar." When missionaries arrived amongst the Algonquin people of North America, the outraged Algonquin called clock time "Captain Clock" because it seemed to command every act for the Christians. The superintendent of one Native American reservation said the first thing he did was to force people to learn time. Utter nonsense. He forced them to learn his time, his clock, his culture's idea of time. To learn that "time is money." To learn coercive, cruel, crushing speed. Punctuality next to godliness. Efficiency *uber alles*.

The British have been particularly good at it. Robinson Crusoe, hero to many, was really a nasty piece of work. Fresh from a failed attempt to be a slave trader, Crusoe took possession of "his" island and imposed a calendar on the days, and in a famous act of power finds a servant to dance to the music of his time, whom Crusoe christens *Friday*. Crusoe was a man of his times, for Britain was furiously slave trading and empire building, insisting that it ruled time, declaring Greenwich Mean Time to be "the" time, Greenwich the center of this maritime empire and ruler of time. Reeking with the language of imperialism and smug with the knowledge that time is power, the chief clock at Greenwich was called the "master" clock and it sent out signals to what were called "slave" clocks at London Bridge and elsewhere.

So this one time, and all the time values which go with it, were imposed on numerous cultures across the world in a widespread and unacknowledged piece of cultural imperialism. Those who seek immoral domination of others have always tied their ambitions to time itself. Hitler's Third Reich was famously intended to last a thousand years. And when I hear the current American administration speak of their "Project for the New American Century," my blood runs cold.

⁓

What's the time? It's a dishonest question. A political question. There are thousands of times, not one. Thousands of cultures around the world with their own calendars, their own times. But one mono-time has worldwide dominance. Western, Christian, manufactured by the Industrial Revolution and molded by Protestantism. Mono-time, mass-produced to go global; Gutenberg's first printing, incidentally, was not a bible but a calendar. And the world's biodiversity of time was set to be crushed.

Just as in wildlands there is *bio*diversity, so too there is *tempo*diversity, a myriad of different times: the slow majesty of an eagle's soar, the tree grazing on the wind for hundreds of years, the sudden leap of a salmon or an insect, tickling the minute.

So take off your watch. It will never tell you the time. Time itself—sensuous, poetic, and diverse—is not found in clockwork.

~

All the Native people I've ever met have known that time—real time—
is articulated in nature, not dumbly enclosed in the clock of the
dominant culture. And they know too something that the dominant
culture forgets: that time is a matter of timing. A sense of real time, free
and uncaged, involves spontaneity rather than scheduling, sensitivity to
a quality of time. Unclockable. The San Bushmen of the Kalahari,
before the eviction from their homeland, said they did not plan when to
hunt but rather said they would, "Wait for the moment to be lucky,"
reading and assessing animal patterns, looking for the "right" time.
Everyone who lives on the land knows that the elastic, chancy, sensitive
times chosen for hunting depend on living things, how the living
moment smells. The timing of social events, meetings, and conversations or pauses in conversations is a skilful affair, for timing is variable
and unpredictable. Time is a subtle element that demands creativity and
improvisation, flexibility, fluidity, and responsiveness. Good timing
demands grace. But the dominant culture, far from respecting those
socially graceful ideas of time, chooses to refer disparagingly to being
"on Mexican time," "on Maori time," "on Indian time." It infuriates me.
It is not indigenous people who lack a sense of time but the dominant
culture who lacks it: lacks alert spontaneity, can't flex, can't dance with
the moment when the moment asks to dance.

The dominant culture doesn't live in the fullness of time, as that
lovely phrase has it, but lives in the emptiness of time. Indigenous cultures, by contrast, understand the fullness of time very well: it is they who
see the future as wildly, incipiently, brightly alive when they look ahead
seven generations before making decisions. It is they who know that the
past is not a dead thing, in line behind you, but rather it is alive, vital in
the vital land. The past is under your feet. "History," Aboriginal writer
Herb Wharton said to me, "History comes up from the land." Which is
why so many indigenous people refuse mining on their territories.

A friend of mine, Sure-yani Poroso, a leader of the Leco people in
Bolivia, told me about his campaigns against mining on his land: "The
land is linked to memory," he said, "so you can't take out the gold and

minerals. They are part of the body of mother earth and we protest against companies destroying our lands. This is a polemic of memory." Sure-yani was tortured for his activism and when it was happening he was told: "You are a little shit. This is what you get for standing against progress." When the word on a torturer's lips is "progress," you see it for the vicious ideology it is. In West Papua, there is an ongoing genocide against the Native peoples. Papuan people said to me: "We are being killed for this idea called 'progress.'" The genocide is carried out by Indonesia, and supported by the British and American government and corporations.

~

Enough of the dark stuff. Let's look to the light. Although wild free time has been caged in clocks, there has been widespread revolt. In Britain, for example, during the Industrial Revolution, the imposition of factory hours meant that common people felt their own time was stolen from them, and in eloquent violence they smashed the clocks above the factory gates that had stolen their time. And the Industrial Revolution has never quite killed the Ludic Revolution: the work ethic has never quite overcome the play ethic.

So let us play. A sense of play, serious play, in Indian mythology is the deepest energy in creation. Traditionally, many indigenous peoples say they do not work for more than four hours a day, which is also the length of time that Bertrand Russell suggested in "In Praise of Idleness." There would then be neither over-employment nor under-employment. He also argues that, "There is far too much work done in the world, and immense harm is caused by the belief that work is virtuous. Leisure, by contrast, is essential to civilization." The play ethic is far more, well, ethical than the work ethic.

Play matters. In play, time is let off the leash, time goes wild in play. Play is the rainbow, is energy, is wicked flirtatiousness, is the helplessly laughing, the leglessly laddered, the God of things which brimmeth over, the pint down the pub, the dew drop overflow of excess, the resplendently unnecessary, and the one too many that make the whole damn thing worthwhile. Play is harvest, is abundance, is generosity, the harvest of pleasure after work, the excess and the gusto, the more than enough, the gift, the spirit of exchange. Take the word giggling. A one word harvest of play's superfluity, its liquid, lovely overindulgence, it has "G"s to spare. "G," the funniest consonant. You want proof? Gnu. Gneed I say more? And it fills the gaps with "I," the quickest, wittiest, lickspittiest, trippiest, and lightesthearted of all the vowels.

All over the world, wild time, public playtime, has been expressed most jubilantly in carnival. Subversive and mischievous, carnival reverses the norms, overturns the usual hierarchies. Unlike dominant time, which does not refer to nature but to cold lifeless numbers, carnival is tied to nature's time. Carnival transforms work time to wild time, upends power structures, and reverses the status quo. It is frequently earthy and sexual. Carnival is vulgar, of the common people. And it is vulgar in another sense— drunken, licentious, loud, and lewd. People really having a wild time.

Carnival emphasizes commonality; customs of common time celebrated by common people on common land. In Britain, a huge number of these customs disappeared as a result of one thing: enclosures, for when rights to common land were lost so were the common carnivals. And just as land was literally fenced off, enclosed, so the spirit of carnival, its spirit of wildness, its broad, unfettered unbounded exuberance, was metaphorically enclosed. Around the world the story was the same. Christian missionaries outlawed carnivals and festivities of other cultures; Native American potlatches banned. Australian Aboriginal corroborees banned. South American traditional dances and festivals banned.

~

The clock is not a synonym for time. It is, if anything, the opposite of time. "There is no clock in the forest," Shakespeare famously said. No clock for sure, but the forest is full of time itself, every pip in every pippin picking the moment to split, every bird knowing when to migrate. All of nature is full of time, every tide heavy with it. Cultures around the world know time is a lived process of nature, for nature shimmers with the poetry of wild time.

There is a scent calendar in the Andaman forests, star diaries for the Kiwi peoples of New Guinea, and in Rajasthan a moment of evening is called "cattle-dust time." One society in Madagascar refers to a moment as "in the frying of a locust," a very quick moment. The English language still remembers time intrinsically connected to nature so I can say I'm going to do something in two shakes of a lamb's tail, and the English language has many terms for "while" as a passage of time, including the arbitrary and sadly obsolete phrase "pissing while."

This is both poetry and politics, and the political opposition between Captain Clock on the one hand and wild time on the other was perhaps best summed up by the leaders of the rebel Zapatistas in Mexico who insisted their time was not the time of the westernized Mexican government. The Zapatistas took their orders from the peasants, and this was a very slow and unschedulable process. "We use time, not the clock. That is what the government doesn't understand." Subcomandante Marcos, in March 2001 in Mexico City spoke to thousands. "*Tlahuica.* We walk time. *Zoque.* We carry much time in our hands. *Raramuri.* Here the dark light, time, and feeling."

What he conjured was time quintessential. Deep, free, and true. Untamed, vivid, and alive. The wild time that is not found in dead clocks and inert calendars, the wild time that is not money but is life itself in ocean tides, in the blood in the womb, in every spirited protest for diversity, in every refusal to let another enslave your time, in the effervescent gusto of carnival, wild time in wild minds, life reveling in rebellion against the clock.

CHAPTER 2

El Carmen—Wilderness in Mexico

"The Bush"

~

JONATHAN BAILEY

South African Passage: Diaries of the Wilderness Leadership School

The Bush ...

The night
moves through the bush
The bird
calls through the bush
The Hyena
laughs through the bush
The man
thinks through the bush
The lion
coughs through the bush

Sound scatters—

Seeps through the bush;
intoxicates the weary mind
weeps through the soul
of days gone by
of days to come.
The soft cocoon
The sharp reality
light softly calls in the east,
And the day bursts through
the bush
always the bush.
God secure me from security,
now and forever.
Amen.

El Carmen
The First Wilderness Designation in Latin America

Armando J. Garcia
Vice President for Development, CEMEX

It's time to think beyond business and profits. We need to think of how business can help protect our biodiversity and the natural resources we depend on every day.

The experts give us a clear message: We are just in time to protect our remaining wild places on Earth. Furthermore, they tell us that these remaining wild places play a critical role for a sustainable world.

CEMEX is a leading global producer and marketer of cement, ready-mix concrete, and aggregate products. We have operations in more than fifty countries and employ more than 50,000 people around the world. Several years ago, we defined in our business policy a guiding principle to work "In harmony with nature." In 1994 we launched our "Sustainable Development" strategy that included a commitment to develop a business culture and to promote good performance in regard to environment, health, and safety concepts.

At CEMEX, we believe that conservation and management of biodiversity are both business and social responsibility issues. We are all part of ecosystems, and ecosystems provide us with important environmental services. Conservation and management of biodiversity issues are strategic, and should be part of our decision-making process. At CEMEX we have a three-part approach:

- Be responsible with the land we manage
- Mitigate biodiversity risks

• Provide proactive leadership and partnerships to make a positive contribution to the global conservation of biodiversity

Since we know that our influence is global, we act with that responsibility and scope in mind. Many of you have seen CEMEX's publications that communicate the values and wonders of nature. Each book details a new conservation strategy. To date, we have presented thirteen titles worldwide, in collaboration with Sierra Madre, Conservation International, the World Wildlife Fund, The World Conservation Union, The WILD Foundation, and others. However important these fine books are and the impact they make, we are even more committed to "hands-on" conservation.

Patricio Robles Gil first took us to El Carmen more than a decade ago, and we formed a unique partnership with his Mexican NGO, Agrupación Sierra Madre. This unique place was a door for CEMEX to get involved in the conservation of the world's biodiversity. It has been our opportunity to show our commitment.

El Carmen is a highly biodiverse, sky-island mountain range that connects two countries in the heart of the Great Chihuahuan Desert. And, such as is the system in Mexico, private landowners play a big part in national and international conservation. It is important to make this contribution positive, because unfortunately this private landowner situation more often brings fragmentation and degradation of the land.

However, at CEMEX we saw land acquisition as a great opportunity to accomplish precisely the opposite: to bring unity and commitment to the future by purchasing the core area of El Carmen. Also, by working to re-establish large landscapes, we guarantee the corridors for wildlife between the different conservation regions to the north, east, and south. Today we own 75,000 hectares inside and outside the Maderas del Carmen Protected Area. We also have another 25,000 hectares that we manage through conservation easements with our neighbors. And there is more land on the way to join this core area.

Five years ago, when we arrived at El Carmen, we knew about the extensive human impacts such as:

- 150 years of overgrazing and loss of wildlife habitat
- Intensive logging of the forest until the 1960s
- Open-pit mining
- Illegal and subsistence hunting for the logging and mining operations involving several hundred people at different times

We knew, then, that the challenge to restore the area was not easy, and represented a long-term commitment. Our first step was to erase as much as possible the human impact on the land, so we started removing domestic livestock, miles of fences, and tons of trash. Then we invited a group of NGOs, respected conservationists from the public and private sectors, and local ranchers to create an advisory board that has provided guidance to our planning and operating process. We started an intensive baseline inventory of flora and fauna, and we began field research with key species such as the Mexican Black Bear. Given that several of the large mammal species that once inhabited these lands have been exterminated, we started an intensive re-wilding process for habitat and wildlife restoration. In our future plans, and once we have viable and growing populations of prey species, we will consider the return of the Mexican Wolf.

One of the most successful programs until now has been the Desert Big Horn Sheep Restorarion Program. That was accomplished with our partners Sierra Madre and Unidos para la Conservación. We constructed a 5,000-hectare breeding facility that today has approximately 150 sheep in it. Earlier this year we released thirty-five sheep from this facility into their historical free range. This action made history because we brought back this important species (highly representative of a true wilderness) that disappeared sixty years ago from Coahuila and Chihuahua.

We believe in alliances as the only way to succeed in the enormous challenge that is in front of us. And that is why we will bring future partners to this initiative. For example, we are working to develop a memorandum of understanding between Agrupación Sierra Madre, Conservation International, BirdLife International, The WILD

Foundation, and CONANP that will allow us collectively to bring together our resources and secure this transboundary conservation corridor.

It is for me a great pleasure on behalf of CEMEX and our partners in this initiative, to declare the first Private Wilderness Area in Mexico and the first designated wilderness in Latin America. For this we have selected one of the most pristine areas in the northern end of El Carmen Mountains, an area that touches the Rio Bravo and connects with Big Bend National Park. With a surface of more than 10,000 hectares (24,000 acres), this unique and spectacular corner will be the first area to have the wilderness designation from the Mexican government.

It gives CEMEX great pleasure to participate in this new chapter in the history of Mexican conservation and in biodiversity conservation around the world.

The El Carmen Wilderness
Part of the El Carmen–Big Bend Conservation Initiative

Patricio Robles Gil
President, Agrupación Sierra Madre

In an extraordinary cooperative effort, we are pleased to announce the El Carmen–Big Bend Conservation Initiative, including the first wilderness designation in Latin America: The El Carmen Wilderness Area. The El Carmen–Big Bend cooperative effort includes participants from the public and private sectors including Agrupación Sierra Madre, Instituto Nacional de Ecología, Unidos Parala Conservacion, Fondo Mexicano Para la Conservatión, Conabio, Pronatura, CEMEX, CONAP (National Protected Areas Commission of Mexico), Museo Maderas del Carmen, A. C. Nature Reserve, CONECO, Conservation International (CI), The WILD Foundation, BirdLife International, The Nature Conservancy, The International League of Conservation Photographers, Texas Parks and Wildlife, National Parks Conservation Association, and Cuenca Los Ojos. The conservation corridor initiative combines many different types of land, all managed separately but with the core value of wilderness preservation. Figure 1 shows the diverse range of management areas linked in the initiative.

The El Carmen–Big Bend area stretches over nearly 10 million acres of land in northern Mexico and Texas, and is home to bighorn sheep, black bear, wild turkey, and mountain lion. The El Carmen Wilderness is a core zone of 80,000 acres, one of the first of several such areas hoped for in the region. This project is one of the leading

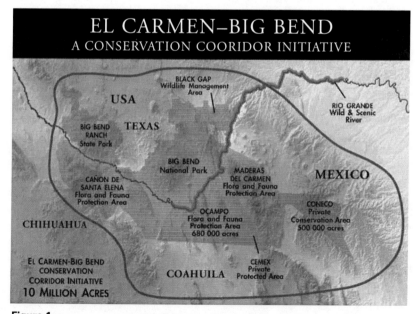

Figure 1

transboundary conservation initiatives in the world, and sets a precedent for effective land designation and management.

Historically, the El Carmen region has drawn considerable attention from the conservation community. Figure 2 shows when specific pieces of the corridor were designated for their specific conservation purpose. In 1999, the Chihuahuan Desert was highlighted by the World Wildlife Fund's "Living Planet" campaign as one of the top twenty priority terrestrial ecoregions. The El Carmen corridor provides many ecosystem services such as carbon sequestration, biodiversity conservation, hydrological services, scenic enjoyment, and economic incentives for private land conservation.

Biologically, the El Carmen region has also drawn a great deal of attention as a "hotspot," with many endemic species and significant habitat loss or risk of loss. In 1998, northern Sierra Madre was listed as a conservation priority for biodiversity as an endemic bird area. As shown in Figure 3, the El Carmen area has a globally significant number of important bird areas (IBAs), meaning that the local bird species fall into one or more of the four IBA categories of globally

Figure 2

IMPORTANT BIRD AREAS

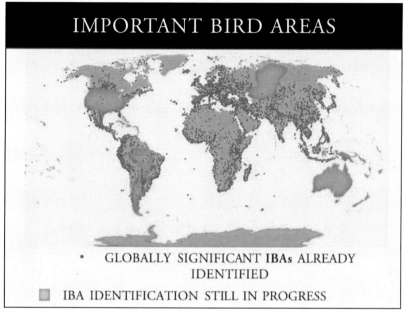

- GLOBALLY SIGNIFICANT **IBAs** ALREADY
  IDENTIFIED
- IBA IDENTIFICATION STILL IN PROGRESS

Figure 3

threatened species, restricted range and biome species, and significant congregatory species.

In 2000, the Mexican National Commission for the Knowledge and Use of Biodiversity identified the El Carmen region as one of terrestrial priority in terms of biological conservation. In 2002, the North American desert region of the El Carmen corridor was highlighted by a wilderness analysis as the fifth most important high biodiversity wilderness area in the world (CI). Also in 2002, the Sky Islands and prairies of northern Mexico were named a priority conservation objective in the "Last Wild Refuges" campaign. Also of regional significance, the Pine Oak Woodlands biome was identified as a new biodiversity hotspot in the 2005 hotspot analysis (CI).

Mexico has several policies in place to encourage private landowners to participate in conservation efforts such as the El Carmen corridor initiative. A volunteer certificate for land conservation recognizes conservation efforts made by private landowners in Mexico. The certificate provides legal protection from Mexico's "idle land legislation" and priority access to economic incentives for the landowners. Mexico has recently introduced a new type of certificate, highlighting the country's commitment to wilderness conservation. The "Wilderness Zone" certificate aims to protect the lands with the highest biological integrity and strengthen the wilderness concept within the Mexican conservation movement.

Protected areas, such as those designated by private landowners using the land conservation certificate, have increased significantly in the five-year span from 2000 to 2005. In Mexico, 2000, there were 42.5 million acres, or 7.9% of the terrestrial surface, protected in 127 areas. In 2005, 46.8 million acres, or 9.7% of the terrestrial surface, were protected in a total of 154 areas. Also in 2005, twenty new protected areas totaling 5 million acres were proposed. This steady increase shows a promising future for conservation efforts in Mexico. Figures 4 and 5 show the respective protected areas for 2000 and 2005.

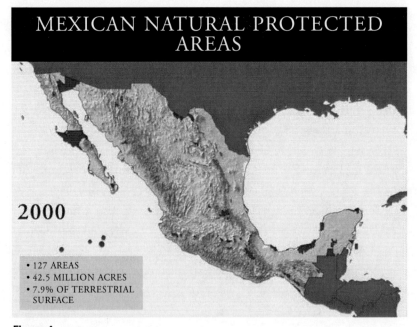

MEXICAN NATURAL PROTECTED AREAS

2000

• 127 AREAS
• 42.5 MILLION ACRES
• 7.9% OF TERRESTRIAL
  SURFACE

Figure 4

These policies and the attention from the conservation community have helped change the El Carmen region from a long history of overuse, including logging operations, mining, cattle ranching, and over hunting. These practices have degraded the land and jeopardized the biological integrity of the region.

In order to reverse the adverse affects of prior abuse and overuse, several re-wildling steps must be taken. These include removing the fences and debris from cattle ranches, completing a baseline inventory of native biology, researching local habitats and ecosystems, and reintroducing wildlife into the area. For example, in regards to the reintroduction of big horn sheep, the Pillares Sheep Reserve established at El Carmen had an inventory of 130 sheep over the five year span from 2000 to 2005 (and in 2004, thirty big horn sheep were found in the wild). Also, in an effort to re-wild the region, an advisory board has been formed to oversee and govern the re-wilding practices.

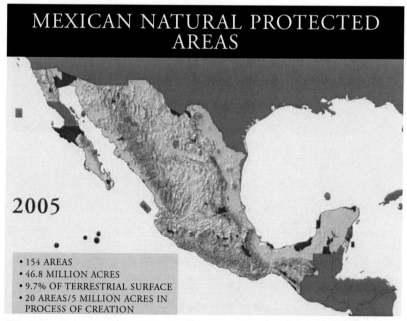

MEXICAN NATURAL PROTECTED AREAS

2005

- 154 AREAS
- 46.8 MILLION ACRES
- 9.7% OF TERRESTRIAL SURFACE
- 20 AREAS/5 MILLION ACRES IN PROCESS OF CREATION

Figure 5

El Carmen is the first wilderness designation in Mexico, and shows commitment to preserve this area of high biological importance, untouched and pristine for the generations to come. It is the first private area in Mexico to be certified as a Wilderness Zone by CONANP (Mexican Commission for Protected Areas), and by a non-governmental coalition as wilderness lands. Both the collaboration and the wilderness and wildlands accomplishment sets a precedent in Latin American nature conservation (Figure 6).

Figure 6

El Carmen—
An Important Conservation Accomplishment

Leon Bennun

Director of Science, Policy, and Information,
BirdLife International

Among many reasons for the significance of the designation of El Carmen as Latin America's first wilderness area, there are three main ones:

First, with biodiversity disappearing faster than ever, and with very limited and inadequate resources to tackle this problem, it is vital that we focus our efforts on the most important places where we can make sure that our investment will have maximum impact.

More than twenty years ago, BirdLife carried out the first study to map global biodiversity using birds. We mapped the distribution of all birds with very small, restricted ranges, less than 50,000 square kilometers. Where these distributions overlapped were "endemic bird areas;" hotspots of endemism and the world's most significant regions for both birds and biodiversity.

El Carmen lies within one of these special places, the Northern Sierra Madre Oriental Endemic Bird Area.

Since then, we have gone a stage further to map out the exact sites, the conservable units, which are the priorities for conservation attention. These are the places that contain the most vulnerable bird species (those that are globally threatened) and those where we have few geographical options for their conservation (such as restricted-range species).

These sites are called "Important Bird Areas" (IBAs). Around 10,000 have been identified so far around the world. El Carmen's special birds, including many migrants and threatened species such as the Black-capped Vireo, qualify it as a globally important bird area.

IBAs can rarely be conserved in isolation; they must be linked together as part of a network. This is exactly what is happening with El Carmen. It forms part of a network of key sites on either side of the U.S.–Mexico border. El Carmen is a key piece in a complex landscape, helping to consolidate the whole and ensure the long-term survival of a suite of remarkable wildlife.

And this brings me to the third reason. Conserving IBAs in the real world is often a complex business, involving different kinds of land management and ownership. Effective partnerships are vitally important. BirdLife itself is a partnership of more than 100 independent conservation NGOs around the world. Partnership is a way of life for us. We draw on each other's strengths and work closely with many other institutions, including the government and private sector, at national, regional, and global levels. It is not always the easiest or quickest way, but it produces the most enduring results.

BirdLife Partners are already working with CEMEX in many countries, including in this endemic bird area, to bring about real conservation gains locally. El Carmen is a great example of the private sector, government, and NGOs working together, in this case to achieve the culmination of a seventy-year-old dream.

CHAPTER 3

Wilderness and Global Biodiversity Conservation

"And the Geese Redeem Me"

~

MARYBETH HOLLEMAN

Winner, 8th World Wilderness Congress Poetry Contest

They waft in, from up and down the greenbelt
which is nothing more than
a small stream running through a big city,
water far from clear,
brambled banks contained by an asphalt bikepath,
graveled playgrounds, baseball fields, parking lots.

Yet all this week,
at the dark end of dusk,
I've walked to where the stream widens into a lagoon,

just to watch them arrive:
waves of Canada geese,
from the manicured lawns of
the oil company complex,
the city golf course,
and thousands upon thousands of yards,
to settle here, on this quiet lens,
appearing with a crescendo of trumpeting calls,
big wings and wide webbed feet angled for the landing,
hovering
over power lines and buildings and down, down onto this circle,
alighting
effortlessly onto this small space amidst an ocean of concrete,
gliding
feather to feather, a thousand or more, on the transformed water,
this one act of redemption each evening,
until ice edges the lagoon and they wing south.

Wilderness, Biodiversity, and the Global Imperative

Michael Hoffmann, Program Officer,
Russell A. Mittermeier, President
Conservation International

We live in a human-dominated planet. During the course of the twentieth century, the global human population increased from 1.65 billion to 6 billion. In April 2005, the results of the Millennium Ecosystem Assessment (MA) were launched globally. The MA, a four-year long assessment involving the expertise of some 1,360 scientists from ninety-five countries and modeled on the Intergovernmental Panel on Climate Change, examined how changes in biodiversity and ecosystem services are affecting people and vice versa. The key finding of the MA is that, over the past fifty years humans have changed these ecosystems more rapidly and more extensively than in any other comparable period of time in human history, largely to meet rapidly growing demands for food, fresh water, timber, fiber, and fuel. The most important direct drivers of change in ecosystems are habitat change, overexploitation, invasive alien species, pollution, and climate change (Millennium Ecosystem Assessment, 2005).

Among the key findings of the MA are:

1. More land was converted to cropland in the thirty years after 1950 than in the 150 years between 1700 and 1850. Cultivated systems, areas where at least 30% of the landscape is in croplands, shifting cultivation, confined livestock production, or freshwater aquaculture, now cover one quarter of Earth's terrestrial surface.

2. In the last twenty years, some regions have experienced very high rates of forest loss, particularly in Central America, Amazonia, the Congo Basin, the forests of eastern Madagascar, and Southeast Asia, mainly in lowland regions. It is projected that a further 10% to 20% of grassland and forestland will be converted between 2000 and 2050, primarily to agriculture. Land conversion is concentrated in low-income countries and dryland regions.

3. More than two thirds of the area of two of the world's fourteen major terrestrial biomes and more than half of the area of four other biomes had been converted by 1990, primarily to agriculture. In contrast, particularly low rates of forest loss were experienced, and are predicated to occur in boreal forests and tundra (Millennium Ecosystem Assessment, 2005).

From a human well-being perspective, such changes are concerning because of the direct linkages that exist between these ecosystems and these essential services that they provide, including: provisioning services (e.g., food, fresh water, fuel), regulating services (e.g., climate regulation, flood regulation, disease regulation), and cultural (e.g., spiritual, recreational, educational). We already know, for example, that the impacts of the massive tsunami that struck Aceh Province in Indonesia on December 26, 2004, could have been significantly less were it not for the loss of coastal mangroves and offshore coral reef systems (Marris, 2005).

These same factors are also leading to a significant, irreversible change in the diversity of life on Earth, and most of these changes represent a loss of biodiversity—the very foundation upon which these ecosystem services are built. We already know that current species extinction rates are 100 to 1,000 times higher than the normal background rate of extinction, estimated at around 0.1 to 1 extinctions per 1,000 species per 1,000 years (Pimm, et al., 1995; Baillie, et al., 2004; Millennium Ecosystem Assessment, 2005). At least in the last 500 years, nearly 800 documented extinctions have occurred, and this is an underestimate being based only on those species known to have gone extinct within this time frame (Baillie, et al., 2004).

Furthermore, the numbers of species threatened with extinction in the near future suggests that we are facing a crisis of unprecedented proportions: according to the IUCN Red List, the global standard for the threat status of species worldwide, one in eight birds, one in four mammals, and one in every three amphibians is at risk of extinction (Baillie, et al., 2004). Of course, again, these are for species that we know. Only 2% of the described 1.9 million species have been assessed using the Red List categories and criteria. The threats to these species include the usual suspects, with habitat loss and degradation the most pervasive threats to species, followed by others such as invasive species, pollution, and over-harvesting. These rates of threat should give us particular cause for concern; amphibians, for instance, are excellent indicators of ecosystem health, so the precarious nature of many populations should serve as a warning for the state of our environment (*see* Stuart, et al., 2004).

Most threatened species occur in the tropics. This is not necessarily surprising given that most biodiversity in general is concentrated in the tropics, with decreasing species richness as one moves away from the poles (Baillie, et al., 2004; Millennium Ecosystem Assessment, 2005). Similarly, threat is not evenly distributed across the face of the planet; rather it tends to be concentrated in particular regions. There are various ways to demonstrate patterns of distribution of threat, including, for example, nightlights, human population density, and maps of net forest loss. Sanderson, et al. (2002) used four types of data, in nine different datasets, as proxies for human influence: population density, land transformation, accessibility, and electrical power infrastructure. The resulting product, the "human footprint," reveals that around 83% of Earth's land surface is currently impacted by human beings. The top 10% of the highest scoring areas look like a list of the world's largest cities: New York, Mexico City, Calcutta, Beijing, Durban, São Paulo, London, and so on. The minimum score (0) is found in large tracts of land in the boreal forests of Canada and Russia, in the desert regions of Africa and Central Australia, in the Arctic tundra, and in the Amazon Basin. The majority of the world (about 60%), however, lies along the continuum between these two extremes, in areas of moderate but variable human influence (Sanderson, et al., 2002).

The uneven distribution of biodiversity and threat creates a dilemma for those of us concerned about the conservation of biodiversity. All biodiversity is important, and we fundamentally cannot accept a policy of triage. But with limited time and resources available, we know we must be strategic about where we act for the benefit of biodiversity conservation. We must identify where we need to act first to prevent massive losses of biodiversity from taking place. Over the last fifteen or twenty years, there has been considerable attention paid to setting global priorities, and also to the theme of systematic conservation planning. The fundamentals of the latter, in particular, hinge on two primary concepts (*see* Margules and Pressey, 2000): irreplacability, which is a measure of spatial options available (in its simplest form, for example, a region having a species confined entirely to within its borders is highly irreplaceable—if it is lost, there are no other options available where that species can be conserved); and vulnerability, which is a measure of temporal options, in others words, how much time we have in which to act (again, in its simplest form, we have less time to act to save biodiversity in a region of higher threat than a region of lower threat). Irreplaceability and vulnerability combine in complex ways that enable conservation planners to identify regions and sites of high priority, with places of high irreplaceability and high threat corresponding to the most urgent priorities.

Such priority regions can be easily mapped into space. Much has been published on biodiversity hotspots (Myers, et al., 2000), those regions characterized by having exceptional endemism (as measured by numbers of endemic plants; 1,500 native vascular plant species) but also under exceptional threat (as measured by percentage habitat loss; at least 70% original native vegetation lost). Currently, some thirty-four such regions are recognized. The habitat within these areas once covered nearly 16% of Earth's terrestrial surface, but remaining habitat now only covers 2.3% of Earth's land surface, as a result of the extensive habitat loss that has taken place within these regions. Nonetheless, even within the habitat that remains, some 50% of all plants and 42% of all terrestrial vertebrates are endemic. More importantly, around three-quarters of the world's most threatened terrestrial vertebrates are found only in the hotspots of biodiversity. At the same time, around one-third of the total human population is found in the hotspots (Mittermeier, et al., 2004).

Another study that echoes the importance of the hotspot regions is that of Rodrigues, et al. (2004a; 2004b), who used five global datasets to determine the current effectiveness of the world's protected areas network, and identified a number of urgent sites where threatened species have no protection whatsoever. These regions show remarkable congruence with the biodiversity hotspots, including: Central America, the Caribbean, the Tropical Andes, Atlantic Forest, Afromontane regions of Africa, Upper Guinea forests, Madagascar, Western Ghats and Sri Lanka, the Himalaya, and parts of Southeast Asia.

However, the problem with conservation action in such regions is that it typically is not cheap: places of high threat tend to be expensive places in which to invest. So what about areas characterized by lower threat? Areas of low threat that generally are more intact usually are cheaper to invest in. But herein lies the paradox: areas of lower threat generally also are of lower biodiversity value. Consider again that these regions, such as the boreal forest and tundra, have undergone the lowest rates of habitat conversion, yet also tend to be poorest in terms of biodiversity value. On the other hand, such areas also are of immense value because of the value they provide in terms of ecosystem services: hydrological control, nitrogen fixation, pollination, carbon sequestration; they are strongholds for large remaining intact faunal assemblages; they represent destinations for ecotourism; they serve as strongholds for many of the world's languages; and they obviously are of great importance for indigenous peoples around the world. Such regions, which we could term "wilderness," offer excellent opportunities for preemptive action at low cost. But, given the uneven spread of biodiversity, how would the goals of such wilderness preservation compare with those of biodiversity conservation?

One way that we can investigate this question is by comparing different approaches to identifying areas of low threat and relative intactness and seeing where and how they differ or overlap, and where and how they compare with regions already known to represent urgent priorities for biodiversity conservation. The first attempt to map wilderness was generated by McCloskey and Spalding (1989), and launched at the 4th World Wilderness Congress. This survey relied on jet navigation charts to identify areas over 400,000 acres (161,943 ha) with no permanent infrastructure. The conclusion was that approximately one-third of the planet consisted of wilderness. In 1994, Lee Hannah and coworkers produced a GIS map of global human disturbance in natural systems. The study produced a habitat index with three categories: undisturbed, partially disturbed, and human dominated. Undisturbed areas retained primary vegetation and had population densities lower than ten people per square kilometer (and under one person per square kilometer for arid, semi-arid, and tundra communities). Partially disturbed areas had secondary but naturally regenerating vegetation with some agricultural development. Human-dominated areas were urban or agricultural. The minimum units mapped were 98,000 acres (40,000 ha). Mixed units were mapped using the dominant landcover, and aggregated into 247,000-acre units (100,000-ha units). The findings were that 52% of the planet was undisturbed, 24% was partially disturbed, and 24% was human dominated.

A third approach to mapping wilderness, developed by James Bryant and colleagues at the World Resources Institute, focuses only on forests and identifies those regions that meet the following criteria (among others): they are primarily forested; they are large enough to support viable populations of all native species, including wide-ranging species (and large enough to do so even in the face of natural disasters); they are dominated by native tree species; and their structure and composition are determined mainly by natural events. The results of this study found that only 22% of the planet's original forest remains as undisturbed "frontier forests" (Bryant, et al., 1997).

Another study, by Eric Sanderson and colleagues at WCS, builds off of the "Human Footprint" project to identify the last of the wild: the 10% wildest areas in each biome in each biogeographic realm. From this set of wildest areas, they selected the ten largest contiguous areas as the "last of the wild," identifying 568 last-of-the-wild areas (Sanderson, et al., 2002).

Finally, Russell Mittermeier and coworkers at Conservation International, attempted to define wilderness from the perspective of a biodiversity conservation organization. They quantitatively defined wilderness regions as those retaining at least 70% of their original habitat, the converse of biodiversity hotspots, and holding human population densities of less than about five people per square kilometer (Mittermeier, et al., 2003). This analysis identified twenty-four wilderness regions and found that while 44% of Earth's land can still be considered wilderness, only five of these regions (their intact portions covering just 6.1% of land) are "high biodiversity wilderness areas" holding as they do more than 1,500 plant species each as endemics. Together, these five regions hold 17% of the planet's plants and 8% of terrestrial vertebrates as endemics.

Overlaying the regions identified as "wilderness" by each of the latter three approaches (the only three to have institutional backing), we can analyze how the goals for wilderness preservation might differ and, more importantly, how the regions of priority differ based on biodiversity. In turn, we can investigate how these regions nestle relative to the biodiversity hotspots and similar areas. Discounting biodiversity value (crudely, by omitting the "high-biodiversity wilderness areas" of Mittermeier, et al., 2003, which is the only approach to directly incorporate irreplaceability; Figure 1; see Brooks, et al., 2006 for methodology), the first clear pattern to emerge is the importance of the boreal forests and tundra regions, as well as the Magellanic forests and the forests of Tasmania (Figure 2). This result is not surprising given the low rates of habitat conversion that have been experienced in these regions. Of course, such an overlay is heavily dependent on the forests-only "frontier forests" analysis; discounting the latter, the most startling pattern to emerge is the agreement on the importance of desert regions, particularly the Sahara, Kalahari, Arabian, Central Asian, and Australian deserts.

If, on the other hand, we factor biodiversity value in, then three regions in particular show up: Amazonia, Congo Basin, and New Guinea (Figure 3). Relative to places such as the biodiversity hotspots, such

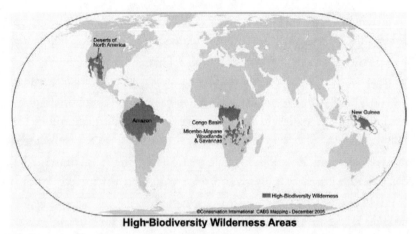

High-Biodiversity Wilderness Areas

Figure 1 High-Biodiversity Wilderness Areas (Mittermeier, et al., 2003)

regions are still relatively intact. However, as the results of the MA attest, these same regions are experiencing, or have experienced, very high rates of net forest loss. These regions, therefore, are not only important priorities because of their biodiversity value, but also because they are facing higher threat and our failure to act now could result in such regions rapidly becoming biodiversity hotspots within a very short space of time.

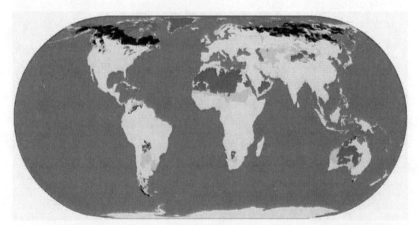

Figure 2 Lower Biodiversity Value. The overlap of regions identified as "wilderness" by Bryant, et al. (1997), Sanderson, et al. (2002), and Mittermeier, et al. (2003), filtered based on biodiversity value using the global biodiversity conservation priority template of Mittermeier, et al. (2003) to illustrate the overlap of approaches within areas of *lower irreplaceability*.

Finally, it is worth remarking on one other means in which the targets for wilderness preservation and biodiversity conservation may not align. In the case of the former, any area protected may possibly suffice; in the case of the latter, however, this absolutely and fundamentally will not suffice. Such ad hoc interventions have already resulted in the incomplete nature of the existing protected areas network (Rodrigues, et al., 2004). Instead, for the purpose of biodiversity conservation, it is necessary for conservation planners to determine strategically where, based on quantitative criteria, important sites are that should be conserved because of their biodiversity attributes. One such example is the Important Bird Areas (IBAs) concept developed by BirdLife International since the early 1980s (Fishpool and Evans, 2001). To date, nearly 10,000 IBAs have been identified worldwide, and efforts are now accelerating to expand the criteria and methodology to capture other biodiversity (under the taxon-neutral Key Biodiversity Areas (KBAs) concept; Eken, et al., 2004). One particularly sensitive subset of these KBAs are those sites holding the last remaining populations of a highly threatened (critically endangered or endangered) species. Such sites, identified by the Alliance for Zero Extinction, represent the most urgent

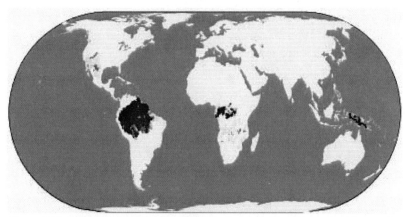

Figure 3 High Biodiversity Value. The overlap of regions identified as "wilderness" by Bryant, et al. (1997), Sanderson, et al. (2002), and Mittermeier, et al. (2003), filtered based on biodiversity value using the global biodiversity conservation priority template of Mittermeier, et al. (2003) to illustrate the overlap of approaches within areas of *high irreplaceability*.

priorities for the conservation of species: places where biodiversity will be lost soon unless we take immediate interventive action (Ricketts, et al., 2005).

In conclusion, wilderness regions have great value to humanity, for the essential ecosystem services that they provide and their benefits in turn for human well-being. However, wilderness preservation alone cannot serve as a surrogate for biodiversity conservation, because of the pattern of distribution of biodiversity, threat, and human influence. Biodiversity conservation must remain firmly targeted on those regions characterized by high threat and high biodiversity value (e.g., Madagascar, Atlantic Forest, Philippines, etc.), and should be complemented by interventions first in those wilderness regions that also are high priorities for biodiversity conservation (in particular, Amazonia, New Guinea, and Congo). Within such regions, any land protected will not suffice. Rather, it is necessary to adhere to a systematic conservation planning framework to determine which parcels of land together will respond to threatening processes and minimize biodiversity loss.

Acknowledgements

Michael Hoffmann and Russell A. Mittermeier would like to acknowledge the work of Vance Martin, Cyril F. Kormos, and John D. Pilgrim as co-authors of this paper.

References

Baillie J.E.M., Bennun L.A., Brooks T.M., Butchart S.H.M., Chanson J.S., Cokeliss Z., Hilton-Taylor C., Hoffmann M., Mace G.M., Mainka S.A., Pollock C.M., Rodrigues A.S.L., Stattersfield A.J. & Stuart S.N. 2004. 2004 IUCN Red List of Threatened Species: A Global Species Assessment. Baillie J.E.M., Hilton-Taylor C. & Stuart S.N., editors. IUCN, Gland, Switzerland and Cambridge, UK. xxiv+191pp.

Brooks T.M., Mittermeier R.A., da Fonseca G.A.B., Gerlach J., Hoffmann M., Lamoreux J.F., Mittermeier C.G., Pilgrim J.D. & Rodrigues A.S.L. 2006. Global biodiversity conservation priorities. Science 313:58–61.

Bryant D., Nielsen D. & Tangley L. 1997. The Last Frontier Forests: Ecosystems and Economies on the Edge. In World Resources Institute, Washington, D.C. Available at: http://www.wri.org/wri/ffi/lff-eng/index.html.

Eken G., Bennun L., Brooks T., Darwall W., Fishpool L., Foster M., Knox D., Langhammer P., Matiku P., Radford E., Salaman P., Sechrest W., Smith M. L., Spector S. & Tordoff A. 2004. Key Biodiversity Areas as site conservation targets. BioScience 54:1110–1118.

Fishpool L.D.C. and Evans M.I. 2001. Important bird areas in Africa and associated islands: Priority sites for conservation. BirdLife Conservation Series No. 11, pp. 1144. Newbury and Cambridge: Pisces Publications and BirdLife International.

Hannah L., Lohse D., Hutchinson C., Carr J.L. & Langkerani A. 1994. A preliminary inventory of human disturbance of world ecosystems. Ambio 23:246–250.

Margules C.R. and Pressey R.L. 2000. Systematic conservation planning. Nature 405:243–253.

Marris E. 2005. Tsunami damage was enhanced by coral theft. Nature 436:1071.

McCloskey M.J. and Spalding H. 1989. A reconnaissance-level inventory of the amount of wilderness remaining in the world. Ambio 8:221–227

Millennium Ecosystem Assessment. 2005. Ecosystems and Human Well-being: Synthesis. Island Press, Washington, D.C.

Mittermeier R.A., Mittermeier C.G., Brooks T.M., Pilgrim J.D., Konstant W.R., da Fonseca G.A.B. & Kormos C. 2003. Wilderness and biodiversity conservation. Proceedings of the National Academy of Sciences of the United States of America 100:10309–10313.

Mittermeier R.A., Robles-Gil P., Hoffmann M., Pilgrim J.D., Brooks T.M., Mittermeier C.G., Lamoreux J.L. & Fonseca G., editors. 2004. Hotspots Revisited: Earth's Biologically Richest and Most Endangered Ecoregions. Second Edition. Cemex, Mexico.

Myers N., Mittermeier R.A., Mittermeier C.G., da Fonseca G.A.B. & Kent J. 2000. Biodiversity hotspots for conservation priorities. Nature 403:853–858.

Pimm S.L., Russell G.J., Gittleman J.L. & Brooks T.M. 1995. Science 269:347–350.

Ricketts T.H., Dinerstein E., Boucher T., Brooks T.M., Butchart S.H.M., Hoffmann M., Lamoreux J.F., Morrison J., Parr M., Pilgrim J.D., Rodrigues A.S.L., Sechrest W., Wallace G.E., Berlin K., Bielby J., Burgess N.D., Church D.R., Cox N., Knox D., Loucks C., Luck G.W., Master L.L., Moore R., Naidoo R., Ridgely R., Schatz G.E., Shire G., Strand H., Wettengel W. & Wikramanayake E. 2005. Pinpointing and preventing imminent extinctions. Proceedings of the National Academy of Sciences 102(51):18497–18501.

Rodrigues A.S.L., Andelman S.J., Bakarr M.I., Boitani L., Brooks T.M., Cowling R.M., Fishpool L.D.C., Fonseca G.A.B., Gaston K.J., Hoffmann M., Long J.S., Marquet P.A., Pilgrim J.D., Pressey R.L., Schipper J., Sechrest W., Stuart S.N., Underhill L.G., Waller R.W., Watts M.E.J. & Yan X. 2004a. Effectiveness of the global protected area network in representing species diversity. Nature 428:640–643.

Rodrigues A.S.L., Akçakaya H.R., Andelman S.J., Bakarr M.I., Boitani L., Brooks T.M., Fishpool L.D.C., da Fonseca, G.A.B., Gaston K.J., Hoffmann M., Long J.S., Marquet P.A., Pilgrim J.D., Pressey R.L., Schipper J., Sechrest W., Stuart S.N., Underhill L.G., Waller R.W., Watts M.E.J. & Yan X. 2004b. Global gap analysis: priorities for expanding the global protected area network. BioScience 54:1092–1100.

Sanderson E.W., Jaiteh M., Levy M.A., Redford K.H., Wannebo A.V. & Woolmer G. 2002. The human footprint and the last of the wild. Bioscience 52:891–904.

Stuart S.N., Chanson J.S., Cox N.A., Young B.E., Rodrigues A.S.L., Fischman D.L. & Waller R.W. 2004. Status and trends of amphibian declines and extinctions worldwide. Science 306:1783–1786.

Wilderness and the Conservation of Biological Diversity

Jeffrey A. McNeely

Chief Scientist, IUCN—World Conservation Union

A Historical Perspective

Our early ancestors were part of the wilderness. Evidence suggests that Africa was our original home, where numerous lines of hominids evolved, diverged, thrived, and died out, leaving *Homo sapiens* at the end of our evolutionary line. It is interesting that the rate of species extinction in Africa is remarkably less than in the other continents, where humans arrived at a later stage in their evolution (Martin and Kline, 1984). In Europe and Asia, early species of *Homo* were found, including *Homo erectus*, *Homo neanderthalensis*, and the recently discovered *Homo floresiensis* (who survived until relatively recently on remote Indonesian islands). All of these used fire and were extremely successful hunters. The evidence suggests that when *Homo sapiens* arrived on new continents, such as Australia and the Americas, numerous species became extinct, and the pre-human wilderness changed its character forever.

Early humans also used fire to hunt, but the result was the extinction of numerous genera of large mammals, such as the wooly mammoth, cave bear, and wooly rhino in the northern hemisphere, and a large number of giant kangaroos and other marsupials in Australia. At later stages, as humans spread into the Pacific, more than 1,000 species of birds were driven to extinction, including the giant flightless birds of New Zealand known as moas (Diamond, 1999; Flannery, 1995).

Hunting and gathering peoples also conserved their resources, often using cultural means such as religion, taboos, or restrictions on hunting and fishing during certain seasons. These helped cultures adapt to virtually all ecosystems on our planet (Suzuki and Knudtson, 1992). Indigenous peoples also developed sophisticated understanding of animal behavior and identified numerous species of medicinal plants. They became an integral part of the wilderness where they lived. Other cultures did not adapt, and died out.

Once agriculture began to develop around 10,000 years ago, the human relationship with wilderness changed fundamentally, and people began to simplify ecosystems to focus on growing relatively few species that were highly productive. As agriculture spread, wilderness was reduced to smaller areas where agriculture could not support dense human populations.

With the development of industrialization about 250 years ago, and greatly accelerating with the discovery of oil as a source of energy in 1859, we underwent another fundamental change in our relation to wilderness. Industrialization and fossil fuels led to rapid expansion of the human population, growing from about 1.6 billion in 1900 to more than 6 billion in 2000. Population growth was accompanied by an impressive growth in the global economy, with gross domestic product (GDP) increasing from about $17 trillion in 1950 to more than $40 trillion in 2004. One effect of growing human dominance has been the growth of cities—today about half the world's population live in cities.

All of these factors have fundamentally changed the relationship among people, wilderness, and biodiversity.

Wilderness, Biodiversity, and the Modern World

Technological change—first with guns, later with trains, highways, and other means—has also driven our changing relationship with the wilderness. For example, off-road vehicles, from four-wheel drive cars and trucks to snowmobiles, have opened up vast areas, often with very damaging effects. The annual Paris to Dakar rally is a sad example of how even the most remote deserts can be torn up, disrupting fragile arid ecosystems that may take decades to recover.

These technological pressures re-emphasize the importance of wilderness as a means of conserving biodiversity. We need wilderness to escape civilization, to experience biodiversity as our ancestors once did. The fact that so many people are appreciating wilderness on foot (albeit with freeze-dried food, GPS, and nylon tents) is an indication of its key function as an antidote to urban society.

The popularity of wilderness as a tourist destination is a relatively recent phenomenon, arising in the past few decades as people became more separated from wilderness. But it also can become an advantage for conservation, as people become reintroduced to wilderness values.

Wilderness also offers numerous benefits beyond the psychological, the most important being the conservation of biological diversity. At a practical level, wilderness provides habitat for wild relatives of domestic plants and animals whose genes are essential to maintaining the food production systems upon which we depend. Wilderness also provides habitat to plants and animals that provide the medicines that maintain our health, or the animals that are hunted to provide sustenance to rural peoples. Wilderness also provides essential habitat to migratory species, and may be the only hope for the world's large predators, essential elements of biodiversity that are unable to survive in close proximity to modern people. Wilderness areas also provide a buffer against disease: as we encroach further into wilderness, we are confronting new pathogens, from Ebola to SARS to AIDS, at increasing rates.

Finally, in addition to mitigating the effects of climate change, wilderness can also help ecosystems adapt to climate change, a critical point as we pour ever more greenhouse gases into the atmosphere. Natural ecosystems containing wilderness areas are far more likely to adapt than are the small areas of isolated habitats (however well protected they may be), and may provide sufficient habitat to enable various species to adapt to changing climate regimes.

Key Problems
Facing Wilderness and Biodiversity

Among the many problems faced by those seeking to promote the biodiversity conservation function of wilderness, some of the most outstanding include:

- Tourism: How can we ensure that the growing numbers of visitors to wilderness areas do not fundamentally change the character of these areas and in fact contribute to their conservation?

- How can we build more support from the public for maintaining relatively large areas of wilderness, and managing such areas in ways that maintain biodiversity?

- How can we best address the eternal conflict between people and nature? While recognizing that many local and indigenous peoples have found ways of living in a sort of harmony with their surroundings, modern consumer society is putting increasing pressure on the wilderness.

- How can we slow the rate of land-use change? Expanding numbers of people are moving into the remaining tropical wilderness areas, such as Amazonia and the Congo Basin. Evidence indicates that forest loss in Amazonia continues to accelerate, exceeding 15,000 square kilometers per year.

- How can we keep invasive alien species out of wilderness areas?

One side effect of globalization has been the rapid movement of some species around the world, leading to the homogenization of biodiversity (Mooney, et al., 2005). The problem of invasive alien species is now recognized as second only to land-use change as a threat to biodiversity, even in wilderness areas.

How to Conserve Wilderness and Biological Diversity

Conservation of wilderness around the world requires multiple approaches, each carefully tailored to the issues of the particular wilderness area. But generally speaking, wilderness can best be conserved by establishing solid links among biodiversity, ecosystem services, and human well-being. The Millennium Ecosystem Assessment recently provided a comprehensive overview of those links (MEA, 2005a; 2005b).

One critical element in this approach is to recognize that wilderness is an expression of culture. Wilderness will continue to exist to the extent that humans ensure that it remains a significant element of our planet. This requires recognition that wilderness conservation is essential for conserving what remains of biodiversity. It also requires awareness on the part of decision makers around the world of the increasing relevance of wilderness in the context of climate change, and the need to increase support to wilderness conservation as a contribution to human well-being as well as biodiversity. A final, though somewhat paradoxical, element is the realization that wilderness areas will require active management to ensure that they can provide the full range of goods and services we desire.

References

Diamond Jared. 1999. Guns, Germs, and Steel: The Fates of Human Societies. W.W. Norton, New York.

Flannery Tim. 1995. The Future Eaters: An Ecological History of the Australasian Lands and People. George Braziller, New York.

Martin Paul S. and Klein Richard G., editors. 1984. Quaternary Extinctions: A Prehistoric Revolution. University of Arizona Press, Tucson.

MEA (Millennium Ecosystem Assessment). 2005a. Ecosystems and Human Well-being: Synthesis. World Resources Institute, Washington, D.C.

MEA (Millennium Ecosystem Assessment). 2005b. Ecosystems and Human Well-being. Biodiversity Synthesis. World Resources Institute, Washington, D.C.

Mooney H.A., Mack R.N., McNeely J.A., Neville L.E., Schei P.J. & Waage J.K.. 2005. Invasive Alien Species: A New Synthesis. Island Press, Washington, D.C.

Suzuki David and Knudtson Peter. 1992. Wisdom of the Elders: Sacred Native Stories of Nature. Bantum Books, New York. 274 pp.

Government Commitments in the "Program of Work on Protected Areas" of the Convention on Biological Diversity

Kalemani Jo Mulongoy, Sarat Babu Gidda, and Marjo Vierros

Secretariat, Convention on Biodviersity

Over the past century, biodiversity loss has taken place at an unprecedented rate. Setting aside areas for special protection has long been used as a way to counter this loss. However, the established protected areas have not always been representative of all the biomes, species, and genetic resources requiring protection, nor managed effectively to respond to their protection objectives.

The Conference of the Parties to the Convention on Biological Diversity adopted, at its seventh meeting in February 2004, a program of work on protected areas (decision VII/28). This program of work is of particular relevance to the theme of the 8th World Wilderness Congress "Wilderness, Wildlands, and People—A Partnership for the Planet," to generate up-to-date information on the benefits of wilderness and wildlands to contemporary and traditional societies, and to review the best models for balancing wilderness and wildlands conservation with human needs.

This paper describes the program of work on protected areas adopted within the Convention on Biological Diversity. After a description of the main objectives of the convention and its programs of work (section two) and the consideration of factors that warrant a program of work on protected areas (section three), the paper presents in section four the overall objectives and contents of the program of work, with emphasis on outcome-oriented targets. Section five highlights references to protected areas in the other programs of work adopted within the Convention on Biological Diversity. In section six, some key issues regarding the way forward are discussed.

The Convention on Biological Diversity and Its Programs of Work

Biological diversity, in short biodiversity, is the variety and variability among living organisms and the ecosystems that support them. Biodiversity is a source of various types of services (provisioning of goods, regulation of ecosystem functioning, cultural services and supporting services that maintain the conditions for life on Earth) important for life on Earth and thus the foundation of human well-being, including security, resiliency, social relations, health, and freedom of choices and actions, upon which civilizations have been built.

Various reports, including in particular the Millennium Ecosystem Assessment (2005), indicate that due to human activities, biodiversity is being lost at an unprecedented rate and that most drivers of change that cause biodiversity loss and lead to changes in ecosystem services are increasing in intensity. Sustaining biodiversity in the face of considerable pressure from human activities therefore constitutes one of the greatest challenges of the modern era. The importance of this challenge was universally acknowledged at the Earth Summit on Environment and Development in Rio de Janeiro, Brazil, in 1992, when the Convention on Biological Diversity (CBD) was adopted to promote the conservation of biological diversity, the sustainable use of its components, and the fair and equitable sharing of benefits arising out of the utilization of genetic resources. The convention entered into force in 1993. It now has 188 parties and thus almost universal membership.

The convention sets out broad commitments by governments to take action at the national and regional levels for the conservation and sustainable use of biological diversity. Since its entry into force, the parties have translated some of the provisions of the convention into a number of tools and a series of programs of work for the major biomes on the planet and other important cross-cutting issues. The Conference of the Parties to the Convention has adopted programs of work on the biodiversity of agricultural lands; inland waters; marine and coastal areas; forest ecosystems; dry and sub-humid lands; mountains; protected areas; incentive measures and traditional knowledge. In general, each program of work establishes a vision and basic principles to guide future work; identifies goals, objectives, and activities; determines potential outputs; and suggests a timetable with specific outcome-oriented targets and means for achieving the outputs.

In addition, the Conference of the Parties to the Convention developed a number of tools to facilitate and streamline implementation of the programs of work. They include the principles and operational guidance for the ecosystem approach; the guidelines for incorporating biodiversity related issues into environmental impact assessment legislation and/or process and in strategic environmental assessment; the Bonn guidelines on access to genetic resources, and the fair and equitable sharing of the benefits arising out of their utilization; the Addis Ababa principles and guidelines for the sustainable use of biodiversity; the guiding principles on invasive alien species; the Akwé: Kon voluntary guidelines for the conduct of cultural, environmental, and social impact assessment regarding developments proposed to take place on, or which are likely to impact on, sacred sites and on lands and waters traditionally occupied or used by indigenous and local communities; and the guidelines on biodiversity and tourism development. The ecosystem approach is the primary framework for action under the convention; it offers a powerful strategy for the integrated management of land, water and living resources that promotes conservation and sustainable use in an equitable way.

At its sixth meeting, in 2002, the Parties to the Convention developed the "Strategic Plan for the Convention," which commits parties to a more effective and coherent implementation of the three objectives of the convention, and to achieve by 2010 a significant reduction of the current rate of biodiversity loss at the global, regional, and national level as a contribution to poverty alleviation and to benefit all life on Earth.

The Program of Work on Protected Areas of the Convention on Biological Diversity

Current Coverage and Representativeness of Protected Areas

For more than a century, countries throughout the world have been setting aside areas for special protection because of their natural beauty and their repository status for spectacular biodiversity. Protected areas are the cornerstones for *in situ* conservation of biological diversity and have long been recognized as a key tool to counter the loss of the world's biodiversity.

Globally, the number of protected areas has been increasing significantly over the last decade, and there are now more than 100,000 protected sites worldwide covering about 12% of Earth's land surface, making them one of Earth's most significant land uses. However, while the number and size of protected areas have been increasing, biological diversity loss continues unabated. The existing global system of protected areas is inadequate in several ways:

1. They are incomplete and do not cover or represent many of the planet's ecoregions, unique sites, and biodiversity hotspots. In particular, recent assessments indicate that conservation of marine and coastal biodiversity, including areas beyond the limits of national jurisdiction, is woefully inadequate, with approximately 0.6% of the world's oceans protected and that freshwater ecosystems are less protected than terrestrial ecosystems.

2. They are not fulfilling their biodiversity conservation objectives.

3. Participation of local and indigenous communities in establishment and management of protected areas is inadequate.

4. Protected areas in developing countries are poorly funded. These issues were discussed at length at the Vth World's Parks Congress held in 2003 in Durban, South Africa, and reviewed in CBD Technical series Number 15.

Role and Effectiveness of Existing Protected Areas

Over the last forty years there has been a paradigm shift in the role of protected areas from national parks and reserves to a broader conceptual and practical approach, including sustainable use areas. Currently, it is recognized that protected areas contribute, besides their conservation function, to human welfare, poverty alleviation, and sustainable development. The goods and services that protected areas provide include, *inter alia*, protection of species and genetic diversity; maintenance of ecosystem services, such as watershed and storm protection; carbon sequestration; products for livelihoods of local people (for example, improvement of fishery and forestry yields); and other socioeconomic benefits, such as in relation to tourism and recreation.

However, many protected areas are ineffective for a number of reasons, including:

1. Insufficient financial and technical resources to develop and implement management plans or lack of trained staff

2. Lack of scientific data and information for management decisions, including information on the impacts of resource use and on the status of biological resources

3. Lack of public support and unwillingness of users to follow management rules, often because users have not been involved in establishing such rules

4. Inadequate commitment to enforcing management rules and regulations

5. Unsustainable use of resources occurring within protected areas, including impacts of human settlement, illegal harvesting, unsustainable tourism, and introduced invasive alien species

6. Contribution to poverty where local people are excluded

7. Impacts from activities in land and sea areas outside the boundaries of protected areas, including pollution and overexploitation

8. Poor governance or lack of clear organizational responsibilities for management and absence of coordination between agencies with responsibilities relevant to protected areas

9. Conflicting objectives of the protected areas.

Overall Objective of the Program of Work on Protected Areas of the Convention on Biological Diversity

It is against this background that the Conference of the Parties to the Convention on Biological Diversity at its seventh meeting in 2002—also taking impetus provided by the Millennium Development Goals, the Plan of Implementation of the World Summit on Sustainable Development (2002), and the Durban Accord and Plan of Action from the Vth World's Parks Congress (2003)—adopted the program of work on protected areas with an overall objective to establish and maintain, by 2010 for terrestrial areas and by 2012 for marine areas, "comprehensive, effectively managed, and ecologically representative systems of protected areas" that collectively will significantly reduce the rate of loss of global biodiversity. Implementation of the program of work on protected areas is expected to contribute to the three objectives of the convention: it's strategic plan, the 2010 biodiversity target, and the poverty alleviation and sustainable development targets of the Millennium Development Goals.

Contents of the Program of Work

The program of work on protected areas consists of four interlinked elements mutually reinforcing and cross-cutting in their implementation. In essence, program element one deals with what protected area systems need to conserve and where. Program elements two and three address the enabling activities that will ensure successful implementation of the other program elements, including issues such as the policy environment, governance, participation, and capacity building. Program element four covers the steps needed for assessing and monitoring the effectiveness of actions taken under program elements one to three. Each program element consists of specific goals, outcome-oriented targets, and related activities. The program of work contains sixteen goals with corresponding targets that set specific dates by which respective goals have to be completed. In many cases the program of work identifies indicators needed for measuring progress towards the goals. A list of activities, ninety-two in total, follow each paired goal and target.

Program Elements

Program element one, "Direct actions for planning, selecting, establishing, strengthening, and managing protected area systems and sites" is in many ways the essence of the program of work. The goals, targets, and activities of this program element taken together define the objectives, nature, and extent of the national protected area systems that will, ultimately, constitute an effective and ecologically representative global network of national and regional protected areas systems. Program element one includes establishing and strengthening national and regional systems of protected areas; integration of protected areas into the larger landscape and seascape, and into various sectors of planning; strengthening collaboration between countries for transboundary protected areas conservation; improving site-based planning and management; and preventing the negative impacts of key threats to protected areas. Achieving goal 1.1 is an essential precondition for achieving the overall objective of the program of work.

Program element two is on, "governance, participation, equity, and benefit sharing." Simply stated, achieving the ultimate goal of the program of work (establishing comprehensive, ecologically representative and effective protected area systems) requires that serious and systematic attention be paid to socioeconomic and institutional matters, not just to biological factors and criteria. This program element includes promoting equity and benefit sharing through increasing the benefits of protected areas for indigenous and local communities, and enhancing the involvement of indigenous and local communities and relevant stakeholders. The central importance for protected areas of governance, participation, equity, and benefit-sharing is underscored by devoting one of the four elements of the program of work to this set of enabling activities.

Program element three, "enabling activities," is about creating an environment that will ensure successful implementation of the other program elements. It includes providing policies and institutional mechanisms; building capacity for the planning, designation, establishment, and management of protected areas; applying appropriate technologies; ensuring financial sustainability; and strengthening communication, education, and public awareness. Program element three provides an umbrella for a number of crucial areas where action is needed to establish the conditions and generate the resources, capacities, and public support to plan, establish, and effectively manage comprehensive, ecologically representative systems of protected areas. Achieving the goals and targets under this program activity clearly requires action by policy and decision makers in many sectors other than protected areas. Policies, laws, and resulting economic incentives in the broader economy are the responsibility of a wide range of government agencies and legislative bodies. In many cases, they can only be changed with strong leadership from senior political leaders.

Program element four, "standards, assessment, and monitoring," includes developing and adopting minimum standards and best practices; evaluating and improving the effectiveness of protected area management; assessing and monitoring protected area status and trends; and ensuring that scientific knowledge contributes to protected area establishment and effectiveness. Program element four addresses the need for parties to put in place systems to assess and monitor the effectiveness of their protected area systems. To do so requires a set of standards and criteria against which to measure the effectiveness of management, a system for evaluating the effectiveness of management interventions, and ongoing monitoring of status and trends of both protected areas themselves and the biodiversity that they contain. In addition, it is widely recognized that scientific knowledge of biodiversity needs to be improved and more widely disseminated to those responsible for protected areas management. Implementing the goals under program element four is therefore essential for determining whether the actions taken under program elements one to three are having their intended impacts, and for allowing for changes in management strategies and actions where that is not the case.

Targets

The program of work on protected areas contains specific time-bound targets primarily organized around national-level actions. The overall target deadline for implementation of the program of work is 2010 for terrestrial and 2012 for marine areas. The Conference of the Parties adopted intermediate targets for many activities with time-bound deadlines of either 2008, 2010/2012, or 2015 in recognition of the fact that many of the goals and targets will require a phased step-by-step approach. The targets are outlined in Table 1 in chronological order. For facilitating the use and implementation of the program of work, the Secretariat of the Convention, in collaboration with the World Commission on Protected Areas of the World Conservation Union, IUCN, and The Nature Conservancy, published an action guide describing the potential steps, case studies, tools, and resources for implementation of the many activities of the program of work.

The program of work on protected areas is a framework within which Parties to the Convention may develop national and regional targets and activities, and implement them in the context of their national priorities, capacities, and needs.

Protected Areas in the Work of the Convention

Protected areas form a central element of the work in the thematic areas and cross-cutting issues addressed by the Conference of the Parties to the Convention on Biological Diversity.

1. Because oceans and seas cover 71% of Earth, the under-representation of marine and coastal ecosystems in the current global protected areas system is particularly alarming. At the same time, global and regional assessments indicate that marine biodiversity globally continues to decline rapidly. For example, coral reefs are highly degraded worldwide, approximately 35% of mangroves have been lost in the last two decades, and historical over-fishing has greatly reduced the abundance of large consumer species, including predatory fish. In addition, there are increasing and urgent concerns about the effects of over-fishing and destructive fishing practices on biodiversity.

    Halting, and perhaps ultimately reversing, this trend presents the global community with a formidable challenge. The seventh meeting of the Conference of the Parties to the Convention on Biological Diversity agreed in 2004 that marine and coastal protected areas are one of the essential tools and approaches in the conservation and sustainable use of marine and coastal biodiversity (decision VII/5 on marine and coastal biological diversity). The Conference of the Parties also agreed that a national framework of marine and coastal protected areas should include a range of levels of protection, encompassing both areas that allow sustainable uses and those that prohibit extractive uses, such as so-called "no-take" areas. The conference further recognized that protected areas alone could not accomplish everything, and that sustainable management practices are needed over the wider marine and coastal environment.

2. In the program of work on the biological diversity of inland water ecosystems (decision VII/4), goal 1.2 calls for the establishment and maintenance of comprehensive, adequate, and representative

systems of protected inland water ecosystems within the framework of integrated catchment/watershed/river basin management.

3. The use and establishment of additional protected areas and the strengthening of measures in existing protected areas are identified as some of the necessary target actions for the implementation of the work program on dry and sub-humid lands (V/23, annex 1, part B, activity 7[a]).

4. The expanded program of work on forest biodiversity, which was adopted in decision VI/22, contains a number of activities related to protected areas. The program of work also calls for work on the role and effectiveness of protected areas.

5. Goals 1.1 and 2.3 of the program of work on mountain biodiversity (decision VII/27) contains provisions on how to plan, establish, and manage protected areas in mountain ecosystems, including the buffer zones of protected areas using, as appropriate, planning or management mechanisms such as ecological, economic, and ecoregional planning/bioregional/hazardous areas zoning, so as to ensure the maintenance of biodiversity, in particular ecosystem integrity. Actions 1.2.5 and 2.3.1, in particular, call for the establishment and strengthening of adequate, effective national, regional, and international networks of mountain protected areas, and the promotion of integrated transboundary cooperation, strategies for sustainable activities on mountain ranges and protected areas.

6. The program of work on article 8(j) on traditional knowledge includes a component on protected areas relating to the management of protected areas by indigenous and local communities (decision VI/10). Specific emphasis is put on the respect of their rights when establishing new protected areas (decisions VII/16).

7. The Guidelines on Biodiversity and Tourism Development, adopted by the world community in decision VII/14 of the Conference of the Parties, include guidelines on how to incorporate sustainable use and equity strategies within and around protected areas.

8. The value of taxonomic data in assisting protected areas site selection is recognized in the program of work for the Global Taxonomic Initiative, contained in decision VI/8. Protected areas are also mentioned in connection with identification, monitoring, indicators, and assessments (decision VI/7) and the Addis Ababa principles and guidelines for sustainable use of biodiversity (decision VII/12).

9. In the Global Strategy for Plant Conservation (annex to decision VI/9), the Conference of the Parties adopted targets 4 and 5, which specify respectively that by 2010 at least 10% of each of the world's ecological regions should be effectively conserved, implying increasing the representation of different ecological regions in protected areas, and increasing the effectiveness of protected areas; and protection of 50% of the most important areas for plant diversity should be assured through effective conservation measures, including protected areas.

Key Issues Regarding the Way Forward
Establishment of a Working Group for Follow-Up Actions
Parties to the Convention on Biological Diversity agreed on a far-reaching and ambitious program of work on protected areas for conserving viable and representative areas of natural ecosystems, habitats, and species to achieve the 2010 biodiversity target and to make significant contribution to the Millennium Development Goals. The program of work is the first global intergovernmental agreement that sets measurable targets and timetables for the world's protected areas and identifies a number of actions for meeting those targets.

When it adopted the program of work, the Conference of the Parties decided to make progress, *inter alia*, in the areas of:

1. The establishment of marine protected areas in marine areas beyond the limits of national jurisdiction

2. Mobilization of adequate and timely financial resources for the implementation of the program of work by developing countries,

particularly in the least developed and the small island developing states amongst them, and in countries with economies in transition with special emphasis on those elements of the program of work requiring early action

3. Development of "tool kits" for the identification, designation, management, monitoring, and evaluation of national and regional systems of protected areas, including ecological networks, ecological corridors, buffer zones, with special regard to indigenous and local communities and stakeholder involvement and benefit sharing mechanisms

For this purpose, the Conference of the Parties established an Ad Hoc Open-ended Working Group on Protected Areas to explore the most appropriate ways forward on these issues.

The first meeting of the Working Group was held in Montecatini, Italy, from June 13 to 17, 2005. The main outcomes of this meeting included:

1. The initiation of work to compile and synthesize existing ecological criteria for future identification of potential sites for protection in marine areas beyond the limits of national jurisdiction, as well as applicable biogeographical classification systems, and recommendations concerning cooperation and coordination among various forums for establishment of marine protected areas

2. Agreement on options for mobilizing financial resources for the implementation of the program of work through a variety of funding mechanisms

3. An updated list of tool kits for implementing the program of work, and identification of areas where more is needed. The report of the meeting is contained in document UNEP/CBD/ WG-PA/1/6. The Working Group is also expected to assess progress in the implementation of the program of work and recommend ways and means to improve implementation.

Marine Protected Areas Beyond the Limits of National Jurisdiction

In order to halt the loss of marine and coastal biodiversity globally, there is a need to rise to the challenge of affording appropriate protection to the 64% of the oceans that are located in areas beyond the limits of national jurisdiction. This area, the global ocean commons, covers 50% of Earth's surface, and is under increasing and acute human threat. Many high-seas ecosystems, such as those associated with cold-water coral reefs and seamounts, have extremely high and unique biodiversity. However, these ecosystems are also vulnerable and fragile, and because of this they are threatened by destructive activities such as deep-sea bottom trawling. The protection of high-seas ecosystems can only be achieved through international and regional cooperation. It can be achieved through the use of tools, such as marine protected areas and through prohibition of destructive practices, such as bottom trawling. The immediate and urgent need to manage risks to marine biodiversity of seamounts and cold-water coral reefs, through, for example the elimination of destructive practices, has been highlighted by the seventh meeting of the Conference of the Parties to the Convention on Biological Diversity, and by a number of other international forums including the Fourth and Fifth Meetings of the United Nations Open-ended Informal Consultative Process on Oceans and the Law of the Sea, third informal consultation of States Parties to the Agreement on Straddling Fish Stocks and Highly Migratory Fish Stocks. The issue of marine protected areas beyond the limits of national jurisdiction is also currently being discussed in the Convention on Biological Diversity context by the Ad Hoc Open-ended Working Group on Protected Areas.

Mobilizing Financial Resources

Establishing and managing protected areas costs money. There are significant running costs to ensure that protected areas are effectively protected, that local communities benefit from them, and that protected area values are maintained in perpetuity. Three separate studies estimated the total annual cost for effective management of the existing protected areas in developing countries ranging from $1.1 billion to $2.5 billion per year/and funding shortfall (total cost minus current funding) between $1 to $1.7 billion per year (Figure 1).

Governments are conscious of these estimated shortfalls and, in adopting the program of work on protected areas, they called for increased financing, including external financial assistance for developing countries and countries with economies in transition. The Conference of the Parties therefore urged parties, other governments, and funding organizations to mobilize as a matter of urgency through different adequate mechanisms and timely financial resources for the implementation of the program of work by developing countries, particularly in the least developed and the small island developing states amongst them, and countries with economies in transition, in accordance with article 20 of the convention, with special emphasis on those elements of the program of work requiring early action (paragraph 9 of decision VII/28). The Conference of the Parties also called on parties and development agencies to integrate protected areas' objectives into their development strategies (paragraph 11 of decision VII/28). In addition, activity 3.4.7 of the program of work calls for the convening of a meeting of the donor agencies to discuss options for mobilizing funding. In this context, potential donor agencies and other relevant organizations met in Montecatini, Italy, on June 20 through June 21, 2005, and identified possible short-term national, regional, and global level options for mobilizing additional funding to developing countries for the implementation of the program of work on protected areas. Partnership among countries and, within countries, among organizations and different sectors was found as one of the most efficient ways to mobilize resources. In this context, a consortium of non-governmental organizations consisting of The Nature Conservancy (TNC), Conservation International (CI), World Wide Fund for Nature (WWF-I), BirdLife International, Wildlife Conservation Society (WCS), Flora and Fauna International (FFI), and the World Resources Institute (WRI) pledged to support the implementation of the program of work at the national level.

The Program of Work on Protected Areas and the World Wilderness Areas

The adoption of the program of work on protected areas by 188 Parties to the Convention on Biological Diversity is of particular relevance to the World Wilderness Congress as a useful framework. The program of work on protected areas offers a unique opportunity for global coordinated actions for the conservation of the world's wilderness areas, especially through implementation of activity 1.1.2 of the program of work, which calls the parties to take action to establish or expand protected areas in "any large, intact or relatively unfragmented or highly replaceable natural areas." Protection of marine wilderness areas can also be undertaken through the establishment of marine protected areas that prohibit extractive activities, as described in section 6.2. Although the primary aim of such marine protected areas is the conservation of biodiversity on the level of ecosystems, species, and genetic resources, they can also provide for sustainable use in the surrounding marine environment, through, for example, the spillover of fish and larvae.

All governments and relevant organizations should take measures to effectively implement the program of work on protected areas and integrate as appropriate its provisions into their strategies, plans, and programs relating to the protection of the world's wilderness. All rele-

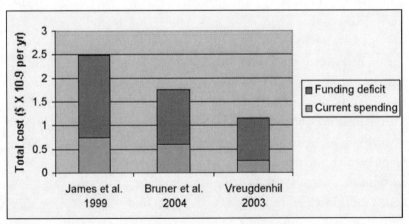

Figure 1 Estimates of funding needs and gaps for effective management of existing protected areas in developing countries (Bruner, et al., 2004)

vant organizations including funding agencies and the private sector should support governments with financial resources and other types of support, in their implementation of the program of work on protected areas of the Convention on Biological Diversity, in particular with regard to the protection of wilderness areas. In this regard effective implementation of resolution 32 of the World Wilderness Congress assumes paramount importance.

Table 1 Targets of the program of work on protected areas

Goal No.	Targets
Targets to be Completed by 2008	
1.5	Effective mechanisms for identifying and preventing and/or mitigating the negative impacts of **key threats** to protected areas are in place
2.1	Establish mechanisms for the **equitable sharing of both costs and benefits** arising from the establishment and management of protected areas
2.2	Full and effective **participation of indigenous and local communities**, in full respect of their rights and recognition of their responsibilities, consistent with national law and applicable international obligations, and the participation of relevant stakeholders, in the management of existing, and the establishment and management of new, protected areas
3.1	Review and revise **policies** as appropriate, including use of social and economic valuation and incentives, to provide a supportive enabling environment for more effective establishment and management of protected areas and protected areas systems
3.4	Sufficient **financial, technical, and other resources** to meet the costs to effectively implement and manage national and regional systems of protected areas are secured, including both from national and international sources, particularly to support the needs of developing countries and countries with economies in transition and small island developing states
3.5	**Public awareness, understanding, and appreciation** of the importance and benefits of protected areas are significantly increased
4.1	**Standards, criteria, and best practices** for planning, selecting, establishing, managing, and governance of national and regional systems of protected areas are developed and adopted *(continued)*

(**Table 1** continued)

Goal No.	Targets
Targets to be Completed by 2010	
1.1	**Terrestrially, a global network** of comprehensive, representative, and effectively managed national and regional protected area system is established
1.3	Establish and strengthen **transboundary protected areas**, other forms of collaboration between neighboring protected areas across national boundaries and regional networks to enhance the conservation and sustainable use of biological diversity, implementing the ecosystem approach and improving international cooperation
3.2	Comprehensive **capacity building** programs and initiatives are implemented to develop knowledge and skills at individual, community, and institutional levels, and raise professional standards
3.3	The development, validation, and transfer of **appropriate technologies and innovative approaches** for the effective management of protected areas is substantially improved, taking into account decisions of the Conference of the Parties on technology transfer and cooperation
4.2	Frameworks for **monitoring, evaluating, and reporting** protected areas management effectiveness at sites, national and regional systems, and transboundary protected area levels adopted and implemented by parties
4.3	National and regional systems are established to enable effective **monitoring** of protected area coverage, status and trends at national, regional, and global scales, and to assist in evaluating progress in meeting global biodiversity targets
Targets to be Completed by 2012	
1.1	In the **marine area**, a **global network** of comprehensive, representative, and effectively managed national and regional protected area system is established
1.4	All protected areas to have **effective management** in existence using participatory and science based site planning processes that incorporate clear biodiversity objectives, targets, management strategies, and monitoring programs drawing upon existing methodologies and a long-term management plan with active stakeholder involvement
Targets to be Completed by 2015	
1.2	All protected areas and protected area systems are integrated into the wider landscape and seascape and relevant sectors, by applying the ecosystem approach and taking into account ecological connectivity and the concept, where appropriate, of ecological networks

CHAPTER 4

The Benefits of Protecting and Sustaining Wilderness

"Lungdhar (Windflag)"

~

KIMBERLEY CORNWALL

Honorable Mention, 8th World Wilderness Congress Poetry Contest

In our final hour, we learn
the lesson of levity. Until then,
a prayer flag shifts in the wind. Maybe longing
is tethered to awe. We are all
flawed students of the same storm.
In peace, in grief, at war—
every hour is a new school.
When there are no headmasters left
our seasons teach each other. Now,

fall berries stain our teeth,
and scars are a Sanskrit
we may never understand.
You rise to the prayer you have,
thread by thread,
disappearing on the saddle of the wind.
The sky has a mission in every direction
and in this place let your lifting begin.

(for Wendy, ascending)

The Global Economic Contribution
of Protected Natural Lands and Wilderness
through Tourism

H. Ken Cordell

Project Leader/Senior Scientist, USDA Forest Service

J. M. Bowker

Research Social Scientist Forestry Sciences Laboratory, USDA Forest Service

This paper reports first-round results of a project aimed at exploring on a global scale the complex relationships between protected natural lands, tourism, and economic growth. In this first round we mainly were interested in secondary sources of data and parameters from previously published studies. In presenting the results we provided summaries of the area of protected natural lands, estimates of the economic impacts stimulated by tourism drawn by these lands, and the spatial distributions of these lands and impacts around the globe. We were surprised to discover that only a very limited amount of research has been done previously to assemble the concepts, models, data, and summaries necessary for such an effort at a global scale. Thus it was necessary for this project to more tightly define the concepts of tourism and nature-based tourism that are relevant to assessing global impacts. Next it was necessary to identify and obtain contemporary data enumerating tourists, their travels, and their spending. Finally it was necessary to pull key concepts and data together for defining, quantifying, and spatially marking the economic activities associated with tourists traveling to visit and see protected natural lands.

Protected Natural Lands

The World Resources Institute (WRI) has listed eight ecosystem types ranging from marine to polar in its recent partnership publication, the Millennium Ecosystem Assessment (Millennium Ecosystem Assessment, 2005). The WRI assessment reported that the structure of the world's ecosystems has changed more rapidly in the second half of the twentieth century than at any other time in human history. The ecosystems that have been most significantly altered globally by human activity include marine and freshwater ecosystems, temperate broadleaf forests, temperate grasslands, Mediterranean forests, and tropical dry forests. Acceleration of human demands have resulted in unsustainable use of natural lands with the result that 60% are now seriously degraded. In an effort to address degradation and conversion to cultivated or developed uses, a number of countries and organizations have been working toward greater protection. Between 1962 and 2003, the world listing of protected natural areas increased from 9,214 to more than 100,000 (United Nations Environment Program, 2003). The area in protected status has risen from 2.4 million in 1962 to 20.3 million square kilometers. Based on the Millennium Assessment, currently about 11.3% of Earth's terrestrial area is now classified as protected. These protected lands are often a draw for tourism.

Tourism

Tourism includes any number of activities that involve persons traveling to and staying in places outside their usual environment. The broad type of tourism most relevant to this paper is nature-based tourism, which includes trips to see, photograph, or visit both protected and unprotected natural lands. Globally, tourism has been growing rapidly. Annually, millions of people travel to see and experience natural lands. In the process of traveling to and/or going into destination natural areas, tourists purchase transportation, lodging, food, souvenirs, and crafts and thus create economic impact. About 11.3% of the natural lands people travel to see are protected. The economic impact of protected-land tourism accounts for the number of travelers, amounts they spend, and how their spending spreads through an area.

Approach

The approach adopted for this paper was to obtain overall measures of the global economic impact of tourism, and then through several steps to disaggregate the relevant economic measures into proportions attributable to the tourism associated with protected natural lands. The World Travel and Tourism Council (WTTC, 2005), and its Oxford Economic Forecasting partner (OEF), have been improving estimates of the economic impact of tourism (Organisation for Economic Co-operation and Development, 2001). Disaggregating their resulting Tourism Satellite Accounts (TSA) was basic to our study as follows:

1. Identify the proportion of total world tourism spending motivated by travel to see and/or visit natural lands

2. Identify the proportion of world nature-based tourism that is attributable to travel to see or visit protected lands

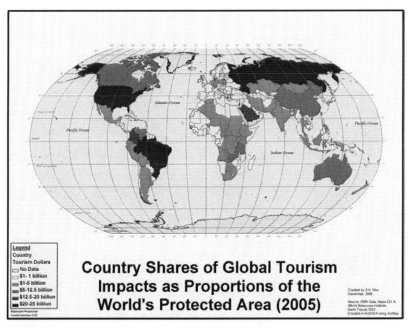

Figure 1

3. Identify the proportion of world nature-based tourism that is attributable to protection of IUCN-World Conservation Union Class 1b, wilderness

4. Identify the proportion of world protected natural lands and wilderness that is attributable to protection in the United States

Results

Total global tourism impact was estimated by the World Travel and Tourism Council at $6,201.5 billion in 2005. The literature indicates that approximately one third of tourism travel is nature motivated. One third of this $6,201.5 billion is $2,066.96 billion, which we considered to be a reasonable estimate of global nature-based tourism impact in 2005. Since 11.3% of global natural lands are protected, it seems reasonable to assume that 11.3%, or $233.6 billion per annum, of nature-based tourism can be attributed to natural lands that are protected. Using this same approach to disaggregation, we estimated that U.S. protected lands contribute $34.6 billion per annum to the U. S. and world economies. Of the U.S. protected lands, 14.3% is designated Wilderness, which we estimate contributed $4.9 billion in 2005. As a cross check, we compared our result with Filion's nature tourism estimate and found it compared very favorably with our estimate of $2,067 billion of global economic impact from nature-based tourism.

The Future

The World Tourism Organization (WTO, 2005) projects tourism will continue growth well into the future. Trips taken for nature-based tourism in the U.S. are projected to grow between 110 and 145% by 2020 for many activities (Bowker, et al., 1999). As the world's human population and economic means grow, unprecedented pressures are being placed on its natural lands as places of interest to see, photograph, visit, and admire. However, the natural attraction of many of these areas as places where tourists can see and experience natural settings is quickly being degraded. Will the economic contribution of protecting areas in the end outweigh the multiple pressures that have led to their continuing demise as natural areas? The world's human population is growing at around 6.2 million per month (U.S. Census Bureau, 2005). In the face of this growth and its migration around the world, it is clear that different biomes in different regions of the world will encounter significant human impacts. With the world's protected and unprotected natural lands contributing more than $2 trillion per year, is it worth adding protection status to more of the world's unprotected natural lands?

References

Bowker J.M., English D.B.K. & Cordell H.K. 1999. Projections of outdoor recreation participation to 2050. In H.K. Cordell, et al. Outdoor recreation in American life: a national assessment of demand and supply trends. Champaign, IL: Sagamore Publishing.

Filion Fern L., Foley James P., Jacquemot Andre J. 1994. Economics of global ecotourism. In Mohan Munasinghe and Jeffrey McNeely, editors. Protected Area Economics and Policy: Linking Conservation and Sustainable Development. Washington, D.C.: World Bank.

Millennium Ecosystem Assessment. 2005. Ecosystems and human well-being: general synthesis. Washington, D.C.: Island Press.

Organisation for Economic Co-operation and Development. 2001. Tourism Satellite Account: Recommended Methodological Framework. Retrieved on December 8, 2005, from http://www.oecd.org/document/27/0,2340,en_2649_34389_1883547_1_1_1_1,00.html.

United National Environment Program. 2003. 2003 United Nations List of Protected Areas. Retrieved on February 23, 2006, from http://www.unep.org/PDF/Un-list-protected-areas.pdf.

U.S. Census Bureau. 2005. World POPClock Projection. Retrieved on December 15, 2005, from http://www.census.gov/ipc/www/popclockworld.html.

World Travel and Tourism Council (WTTC). 2005. Executive Summary, Travel and Tourism: Sowing the Seeds of Growth. Retrieved on December 8, 2005, from http://www.wttc.org/2005tsa/pdf/Executive%20Summary%202005.pdf.

Umzi Wethu
Nature, Nurture, Future

Andrew Muir

Executive Director, The Wilderness Foundation South Africa

"We see wilderness as a force for social change. It has the power
to transform. We cannot ignore the human crisis of HIV/AIDS
or avoid the story history tells of wilderness repeatedly
destroyed during times of human conflict."

—Andrew Muir

The shattering impact of HIV/AIDS increases daily. As of October
2005, there are more than 900,000 orphans and vulnerable youth in
South Africa. Most of these victims face overwhelming challenges;
young people forced to head households before their time. Intervention
must be direct, supportive, and must enable them to redefine a future.
Investing in these young people is far-reaching because they are crucial
to entire family systems.

Welcome to the vision of Umzi Wethu, a project harnessing the
power of nature to secure a future for displaced young people. Umzi
Wethu is a safe and supportive home environment for specialized skills
training with agreed placement into a job in the booming ecotourism and
hospitality industries (and, in time, other industries). It focuses on young
people orphaned due to the HIV/AIDS pandemic, who would otherwise
be unemployed.

An Umzi Wethu Training Centre combines mentoring with the
transformative power of wild areas, alongside life skills and accredited
hospitality and/or conservation knowledge. Learners receive constant

intervention for a minimum of eighteen months in a mix of natural environments and a home campus environment. Every participant spends at least 20% of their training time in wilderness and wild areas.

With a broad network of partnerships, Umzi Wethu provides young people with a future through crucial job creation and its multiplier effect. It creates a bridge beyond existing orphan support lines, introducing a strong underlying environmental ethic.

Partners and organizations that have developed and designed the project to date include: African Global Skills Academy; Barnabas Trust, Cape Action Plan for People and Environment; Eastern Cape Training Centre; Eastern Province Child and Youth Centre; Eastern Cape Private Game Reserves (Indalo); Endangered Wildlife Trust; Hope Worldwide; Hope Africa; Nelson Mandela Metropolitan University (NMMU) Health Clinic; NMMU Architectural Department; SOS Children's Villages; Ubuntu Education Fund; Wilderness Foundation South Africa and The WILD Foundation.

A strategic blueprint has been developed with the minds of experienced practitioners from more than thirty partner organizations. A key objective of the program is to develop and maintain committed

Umzi Wethu trainees spend at least 20% of their time on trail in the wilderness, learning the skills and feeling the spirit of the bush.
Photo by Debbie Gothan

partnerships to establish and ensure the sustainability of the Umzi Wethu program.

The founding Umzi Wethu Academy is based in Port Elizabeth in the Eastern Cape, and project managed by the Wilderness Foundation. The first participants for the day program were selected in March 2006, and began their skill-based training in May 2006.

It has been agreed that Umzi Wethu academies can be both urban and rural based. This allows for both the use of existing infrastructure and support services in an urban setting, and for child headed household to participate without causing a negative disruption to the household.

South Africa is a world conservation leader, and celebrates ten years of democracy. But beneath this huge success grows social devastation of proportions that will overwhelm South Africa's capability to sustain its achievements. Umzi Wethu is a proactive approach, giving displaced teenagers the opportunity to change their destiny from a life of poverty, and possibly crime, to economic advancement and social leadership embracing environmental values.

A few of the trainees in the Umzi Wethu Academy for Vulnerable Children, during the catering course. Photo by Vance G. Martin

Imbewu
Connecting Culture and Conservation

Fanyana Shiburi
Trustee, Wilderness Foundation, South Africa

Wilderness is an enigma; it is different things to different people. Generally there are two main camps—the scientists and the non-scientists. While scientists need empirical evidence and the facts about wilderness to validate their theories, the non-scientists need to be in wilderness to fully understand (and never understand) its power and connectivity. I like to call these people the spiritualists, or poets.

Ian McCullum says in his new book titled *Ecological Intelligence*:

"When we no longer shudder at the ecological warning calls of science, it would seem that the only voice left that can awaken us belongs to the poets. Poetry comes at us from both sides, from inside and from out. It will not let us off the hook and if we listen to the language carefully, it should not take long to understand that it is the language of soul. We have to be able to shudder."

This paper will discuss a South African wilderness and social intervention program called "Imbewu." Ian Player's book *Zulu Wilderness* is what Imbewu is, a grappling of the soul within wilderness and within humanity.

The Imbewu Program in South Africa unashamedly emanated from the poetic or spiritual view of wilderness, and has maintained this core philosophy for the last nine years. We need no empirical proof of the effects of wilderness on the human psyche. We know in ourselves, we trust, that nature has in it a profound healing power and we go there to start that process. We do not believe that Earth needs healing; that is an anthropocentric and outdated assumption! It is we who need healing, so that we can stop our wanton destruction of Earth. The very thing that humanity is subconsciously bent on eradicating (nature) is our very salvation.

Indigenous people—whether they are the San of southern Africa, the Indians of the Amazon, or the Inuit of Alaska—have always been connected into Earth's system and have understood the reciprocity that is required. Western culture has taken us away from that. But through Imbewu we are trying to re-understand.

The essence of the Imbewu Program is a reawakening of ancient cultural knowing that connects us to nature and nature to us. It is re-establishing that link between nature and culture that makes us self-aware and not selfish, where we regain our self-identity and therewith our self-esteem and can harness this new confidence to empower us to create a sustainable world. We believe that no skills development, capacity building, good governance is possible without this fundamental learning.

The Columbian Indians believe that to solve a problem one has to go back to the beginning. There is no better beginning than the healing of the human psyche or spirit, disrupted by the negative effects of colonialism and apartheid in South Africa, or by many other things in other parts of the world. One cannot build a skyscraper on a bed of clay, and this is what we are trying to do if we do not look first to ourselves.

We have put these two concepts of the connection between nature and culture in the worldview of First Peoples and the healing powers of wilderness together and created a program that benefits conservation, culture, and communities. It is a public/private partnership between South African National Parks and the Wilderness Foundation of South Africa, and is run in the game and nature reserves in five of the nine provinces. Our aim is to increase it to seven provinces by the end of next year.

Imbewu is led by elders from the various cultural groups who have been retired from the national parks, having been game guards for most of their lives. They have an abundance of knowledge gained both from their own elders, as they were raised in the traditional way, and from their years of work quietly listening to what was going on around them. The recognition of the value of their knowledge in this new age has returned to them a strong sense of self-esteem and raised their standing in their communities where traditional values are disintegrating and established them as role models to the youth in general. Since Imbewu's inception in 1996, 6,000 youth have passed through the program.

Imbewu identifies potential leaders of tomorrow, between the ages of sixteen and twenty, from previously disadvantaged schools and takes them into the wilderness for four days and three nights to interact with the elders and sleep under the stars. Facilitation is done through the traditional method of storytelling and anecdotes in the ethnic language of the area. The days are passed with interpretive hikes through the bush, camp activities, solitary time in a beautiful space, and preparing meals. After supper, storytelling around the campfire often leads to animated discussion.

The element that has the greatest reported impact is the solitary time around the fire on night watch, as the trust of the whole group is placed in the hands of one person for one hour each. As the learners come face to face with their fears of the perceived dangers of the bush and the realization that many of our realities are created by incorrect perceptions, a deeper sense of our connectedness and our ability to create a new reality emerges. Not only is a better understanding developed, but personal growth is also enhanced.

Imbewu was designed as an entry-level experience into wilderness. We are now realizing that it has the potential to be far more than this. Using participatory methodologies we are now redesigning the program to give the learners more support once they leave the wilderness to establish micro projects in their schools and communities that satisfy community needs and truly empower them as leaders of tomorrow.

An exciting association with the African Biodiversity Network has meant that the Imbewu Program has crossed the boarders of South Africa and has became Imbewu Pan Africa, now running in Ethiopia and as a part of the expansion of the Green Belt Movement in Kenya.

But reciprocity applies here, too. South Africa hosted representatives from these two countries on a month-long experiential learning program, visiting all five of the Imbewu camps around the country and participating in various workshops. I defy anyone to claim that they have traveled more kilometers around South Africa than they! A delegation of Kenyans and South Africans spent two weeks in Ethiopia learning from the Cultural Biodiversity Program in schools there. This has the potential to be the core of the new school and community element of Imbewu Projects in South Africa. These mutual learning experiences are very exciting, as it is African countries sharing with each other with an environmental integrity that is uniquely African.

We don't need empirical evidence to tell us the effect that Imbewu has on the lives of these children. We just have to watch their faces as they open themselves to nature's soul over the course of four days.

The Economic Value of Ecosystem Services from and for Wilderness

Trista Patterson
PNW Research Station, U.S. Forest Service

In the Sierra Club classic *On the Loose*, brothers Terry and Renny Russell reject attempts to place economic values on wilderness, emphasizing that the true rewards of the wilderness experience are spiritual:

the freedom of self-reliance and the uplifting beauty of wild nature. At the same time, citing Winston Churchill, they issue a key challenge: to learn the game one has to play, for more than one can afford to lose. Some wilderness scholars are taking up this dare by re-examining and re-employing economic tools they had long since dismissed. Economic valuations of wilderness have concentrated especially on direct (e.g., commodity goods, recreation) and non-use (e.g., existence, bequest) benefits (Haynes and Horne, 1997; Schuster, et al., 2005; Cordell, et al., 1998; Loomis, 2000; Loomis and Walsh, 1992; Loomis and Richardson, 2001; Richardson, 2002; Walsh, et al., 1984; Walsh and Loomis, 1989). Increasing public importance has been noted for *indirect* values from wilderness, such as ecosystem services (Figure 1; Morton, 1999, 2000; Cordell, 2003). Ecosystem services are the naturally occurring contributions to life-support, and quality of life that people normally don't have to pay for[1] (Daily, 1997; Costanza, et al.,

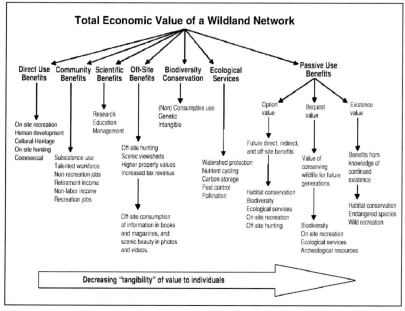

Figure 1 Morton's (2000) total economic valuation framework for estimating wilderness benefits based on seven categories arranged from left to right in order of decreasing tangibility to humans.

1997; de Groot, et al., 2002). They can be experienced directly—provisioning food, fiber, freshwater, and cultural and recreational opportunities—or indirectly—regulating floods or climate or supporting the other services through soil formation, nutrients, and habitat (MA, 2005; Chapin this issue).

Creative experiments are bringing values of ecosystem services into the marketplace, including carbon markets; wetland and habitat banking; water temperature credits; and certifications and tax incentives (Wunder, 2005). Market values have helped raise awareness for ecosystem service contributions to quality of life, and help harness funds for their protection. Achieving these outcomes for wilderness involves particular challenges, four of these follow.

Broadening the Methods

One challenge to economic valuation of ecosystem services from wilderness is that reducing a multi-faceted issue to the market is by nature an exclusionary process (Funtowicz and Ravetz, 1994; Funtowicz, et al., 1999). As a consequence, only a subset of value, population, and worldview is reflected. When Costanza, et al. (1997) estimated the value of the world's ecosystem services as $33 trillion, 1.8 times the world's GDP, some logically wondered how people's willingness to pay could exceed more than they had (Bockstael, et al., 2000). The over-reliance on certain methodologies can obscure worldviews relevant to wilderness; for example, the possibility that the value of the commons could be greater than the sum total of all the things we own as individuals. In addition to neoclassical economic tools, social science deliberative and consensus methods, multi-criteria and conjoint analysis, and ecological pricing (e.g., emergy and exergy) can elucidate and convey values from multiple perspectives (Patterson, 2005). These are necessary to relating willingness-to-pay to the market, the market to the economy, and the economy to wilderness.

Distinguishing Growth from Development

The term economic growth is often used interchangeably with economic development (Daly, 1977), but with different implications for wilderness (Czech, 2000). Growth, a quantitative attribute, involves increasing economic activity—commonly a result of increasing population and/or per-capita energy/material consumption. Technology often does not fully mitigate the impacts of this, or we allow negative consequences to be borne out in future generations. Growing land areas and use intensity needed to support economic growth can ultimately compete with, or adversely impact, wilderness. This occurs not only at geographic boundaries (White, et al., 2000), but also systemic changes in climate, species dynamics, and soil and water transport. In contrast, development, a qualitative attribute, can be achieved by economic rearrangement, in theory improving the ability of wilderness and the man-made economy to coincide. This must be the center of our focus if economic tools are to be harnessed effectively from and for wilderness. Accounting ecosystem services from wilderness can help to distinguish these qualitative improvements.

Developing Creative Markets, Flexible Institutions

Characteristics of goods and services affect the ease of using market based tools to elicit their value. Marketed goods are most often excludable, a legal concept that allows an owner to prevent another person from using the asset, and rival, where consumption or use reduces the amount available for other people, while most ecosystem services are non-excludable, and non-rival (see Daly and Farley, 2004 for applications). To make ecosystem services marketable, to some extent, social agreements can engineer excludability or rivalness, or create a proxy that is (consider carbon credits). Wilderness (often on public land) requires additional creativity because most market-based mechanisms are salient to private lands. That said, offsets elsewhere can benefit the wildland network as a whole, and ecosystem services that are not marketable (e.g., biodiversity) can be bundled to one that is (e.g., water temperature credits).

Regulations, laws and standards, market incentives, information (e.g., certification), and the institutional flexibility all influence the longer standing success of attempts to bring wilderness attributes to market. Simply because the market is trading carbon credits in quantity doesn't mean abatement is occurring. Market price for carbon was more than halved in April 2006, when it was revealed that European countries had set first-round emission targets too high.

Cultivating Socially and Environmentally Just Markets

Links between wilderness and ecosystem services often involve broad spatial scales that are rarely congruent with market and property bounds. Time lags and feedback loops can muddle the cause-effect relations needed to reflect marginal gains. Wilderness affects ecosystem services and vice versa; forest loss in Amazonia reduces rainfall in Texas (Avvisar and Werthe, 2005), while carbon emissions from cities affect Arctic wilderness (Bachelet, 2005).

Conditions that satisfy market efficiency don't include environmental sustainability or socially just distribution[2] (Daly and Farley, 2004). For the world's poorest, ecosystem services provide "natural insurance" for people living in or near wilderness, as has been documented in Peru, the Amazon (Takasaki, et al., 2004), Knuckles Wilderness in Sri Lanka (Gunatilake, 1993), and others (Pattanayak and Sills, 2001). Despite this, wilderness has at times been cast as elitist because demographic disparities exist in those who access it (Johnson, 2004). Exclusive focus on direct, rather than indirect or non-use, benefits can obscure important distributive justice benefits of wilderness.

Conclusion

Wilderness contributes to indirect economic value through broad-scale ecosystem services, buffering severity and directionality of environmental change, and by helping us understand the way nature works. One barrier to stemming the losses of ecosystem services and wilderness alike is an inability to account for their non-monetary contributions to quality of life, or the damage costs to be incurred when they are lost.

Broadening assessment of value to include the indirect (public) goods and services can prevent assets of "the commons" from taking a back seat to private profit, *sensu* Hardin (1968). I've mentioned four challenges particular to wilderness: vigilance that the market and willingness to pay is not the only way we elucidate economic value; distinguishing economic growth a quantitative goal, from economic development a qualitative goal; employing creativity and skill with economic instruments and flexibility with social institutions; and lastly, looking beyond market efficiency to social and environmental justice issues.

The economic approach is not for everyone. If the Russell brothers had been asked to put a dollar value on wilderness, they probably would've responded with a public mooning. Yet the market is already valuing wilderness by way of a very few commodified and direct-use values. Progress from and for wilderness is perhaps most hindered when we don't have any new or compelling tools with which to construct a vision for the future. More use can be made of economic instruments without eclipsing values in social, cultural, or ecological terms.

References

Alcamo J., Ash N.J., Butler C.D., Callicott J.B., Capistrano D., Carpenter S.R., et al. 2003. Ecosystems and human well-being: a framework for assessment. Washington, D.C.: Island Press.

Avissar R. and Werth D. 2005. Global hydroclimatological teleconnections resulting from Tropical deforestation, Journal of Hydrometerorology 6:134–145.

Bachelet D., Lenihan J., Neilson R., Drapek R. & Kittel T. 2005. Simulating the response of natural ecosystems and their fire regimes to climatic variability in Alaska. Can. J. For. Res. 35:2244–2257

Bockstael N., Freeman A., Kopp R., Portney P. & Smith V. 2000. "On Measuring Economic Values for Nature," Environmental Science and Technology, 34(8), 1384–1389.

Boyd J.W. and Banzhaf H.S. 2005. Ecosystem services and government: the need for a new way of judging nature's value. Resources, 158, 16–19.

Brown T., Loomis J. & Bergstrom J. 2006. Ecosystem Goods and Services: Definition, Valuation and Provision USDA Forest Service, RMRS-RWU-4851 Discussion Paper.

Cordell K., Tarrant M. & Green G. 2003. Is the public Viewpoint of Wilderness Shifting? International Journal of Wilderness 9(2)27–32.

Cordell H.K., Tarrant M.A., McDonald B.L. & Bergstrom, J.C. 1998. How the public views wilderness. International Journal of Wilderness 4(3):28–31.

Costanza R., d'Arge R., de Groot R., Farber S., Grasso M., Hannon B., Limburg K., Naeem S., O'Neill R., Paruelo J., Raskin R., Sutton P. & van den Belt M. 1997. The value of the world's ecosystem services and natural capital. Nature, Vol. 387: pp 253–260.

Czech B. 2000. Economic growth, ecological economics, and wilderness preservation. In McCool Stephen F, Cole David N., Borrie William T, O'Loughlin Jennifer, comps. 2000. Wilderness science in a time of change conference—Volume 2: Proceedings RMRS-P-15-VOL-2. Ogden, UT: U.S. Department of Agriculture, Forest Service, Rocky Mountain Research Station. p. 194–200.

Daily G.C. 1997. Introduction: what are ecosystem services? In G.C. Daily, editor, Natures services: societal dependence on natural ecosystems. Washington, D.C.: Island Press, p. 1–10.

Daly H. 1977. Steady state economics. Washington, D.C.: Island Press, p. 318.

Daly H., Farley J. 2004. Ecological Economics, Principles and Applications. Washington, D.C.: Island Press, p. 449.

de Groot R., Wilson M.A. & Boumans R.M.J. 2002. A typology for classification, description and valuation for ecosystem functions, goods and services, Ecological Economics, 41 pp. 393–408.

Funtowicz S. and Ravetz J.R. 1994. The Worth of a Songbird: Ecological Economics as a Post-normal Science, Ecological Economics, 10(3):197–207.

Funtowicz S., Martinez-Aler J., Munda G. & Ravetz J.R. 1999: "Information tools for environmental policy under conditions of complexity." Environmental Issues Series 9, EEA, Copenhagen.

Gunatilake H., Senaratne D. & Abeygunawardena P. 1993. Role of non-timber forest products in the economy of the peripheral communities of Knuckes National Wilderness Area of Sri Lanka: a farming systems approach, Economic Botany 47:25–35.

Hardin G. 1968. The Tragedy of the Commons. Science, 162:1243–1248.

Haynes R.W. and Horne A.L. 1997. Economic Assessment of the Basin. In Quigley T.M., Arbelbide S.J., editors. An assessment of ecosystem components in the Interior Columbia Basin and portions of the Klamath and Great Basins: Volume IV. 1715–1870. USDA Forest Service, PNW-GTR-405, Pacific Northwest Research Station, Portland, OR.

Heal G.M., Barbier E.B., Boyle K.J., Covich A.P., Gloss S.P., Hershner C.H., et al. 2005. Valuing ecosystem services: toward better environmental decision-making. Washington, D.C.: National Academy Press.

Johnson Cassandra Y., Bowker J.M., Bergstrom John C. & Cordell H. Ken. 2004. Wilderness values in America: Does immigrant status or ethnicity matter? Society and Natural Resources 17:611–628.

Landres P.B., Marsh S., Merigliano L., Ritter D. & Norman A. 1998. Ecological effects of administrative boundaries. In: Knight R.L., Landres P.B., editors, Stewardship across boundaries. Washington, D.C.: Island Press, 117:139.

Loomis J., 2000. Economic values of wilderness recreation and passive use: what we think we know at the turn of the 21st century. In Cole David N., McCool Stephen F. 2000. Proceedings: Wilderness Science in a Time of Change. Proc. RMRS-P-000. Ogden, UT: U.S. Department of Agriculture, Forest Service, Rocky Mountain Research Station.

Loomis J., Richardson R. 2001. Economic Values of the U.S. Wilderness System: Research Evidence to Date and Questions of the Future, International Journal of Wilderness 7(1)31–34.

Loomis J., Walsh R. 1992. Future economic values of wilderness. In Payne C., et al., editors. The economic value of wilderness. Southeastern Forest Experiment Station. GTR SE-78.

Millennium Ecosystem Assessment. 2005. Ecosystems & Human Well-being: Synthesis. Washington, D.C.: Island Press.

Morton P. 1999, The Economic Benefits of Wilderness: Theory and Practice. University of Denver Law Review, V 76, No.2

Morton P. 2000, Wildland Economics: Theory and Practice. USDA Forest Service Proceedings RMRS-P-15-Vol-2.2000. In McCool S., Cole D., Borrie W., Loughlin J.O. Wilderness within the context of larger systems. 1999. May, 23–37; Missoula, MT. Proceedings. Rocky Mountain Station. USDA Forest Service Proceedings RMRS-P-15-Vol-2.2000. Ogden, UT.

Pattanayak S. and Sills E. 2001. Do tropical forests provide natural insurance? The microeconomics of non-timber forest product collection in the Brazilian Amazon, Land Economics 77:595–612.

Patterson M. 2005. Valuation in ecological economics: is there and should there be a common ground? Presentation: Ecological Economics in Action, December 11–13, 2005 Massey University, Palmerston North, New Zealand.

Richardson R. 2002. Economic Benefits of Wildlands in the Eastern Sierra Nevada Region of California. The Wilderness Society, San Francisco, CA.

Russell T. and Russell R. 1967. On The Loose. Sierra Club, Ballantine, New York, NY.

Schuster R., Cordell H.K. & Phillips B. 2005. Measurement of direct-use wilderness values: a qualitative study. Proceedings of the 2005 Northeastern Recreation Research Symposium USDA Forest Service GTR-NE-341 p188–196.

White P., Harrod J., Walker J. & Jentsch A. 2000. Disturbance, scale and boundary in Wilderness Management. In USDA Forest Service Proceedings RMRS-P-15-Vol-2. 2000, p. 27–42.

Wunder S. 2005. Payments for environmental services: some nuts and bolts. CIFOR discussion paper. Center for International Forestry Research; Jakarta, Indonesia.

Walsh R.G., Loomis J.B. & Gillman R.S. 1984. Valuing option, existence and bequest demands for wilderness. Land Economics 60(1):14–29.

Walsh R.G. and Loomis J.B. 1989. The non-traditional public valuation (option, bequest, existence) of wilderness. In Freilich H.R., editor, Wilderness benchmark 1988: Proceeding of the national wilderness colloquium. USDA Forest Service GTR SE-51, Southeastern Forest Experiment Station, Asheville, NC.

CHAPTER 5

Water and Wilderness

Freshwater, Oceans, and Wild Seas

"Mist and Heron, Heron and Mist"

~

ANN CHANDONNET

Honorable Mention,
8th World Wilderness Congress Poetry Contest

Grass enameled with jade dew,
And trees reflected above the lake
In scintillating tendrils of white mist.

Audubon's white heron,
Coiled toward the frog
Who leaps off to the left,
Spread flat as a flying squirrel,
mirrors the shape of this mist.

The frog leaps left to safety,
While the feathers fall, curving,
Toward the reeds curving on the right.
These things point down: neck, frog, reeds, plumes.
But the long legs draw the eye up:
And the mist rises up—
An insubstantial fountain.

Form and ritual, an artful psychotherapy,
Ceremony can heal,
Gently,
Like a balm of mist on eyelids,
Like a wash of dew on reeds.

The Last Wild Waters
A Snapshot of the Remaining Freshwater Wilderness

Michael L. Smith
Scientist, Freshwater Biodiversity,
Conservation International

Fresh water is the most precious medium for the conservation of species. Although freshwater habitats cover only 0.8% of Earth's surface, they contain, for example, 31% of the world's vertebrate species (based on an original count of species that live in fresh water at some stage in their life cycle). If the numbers of fish species in fresh and marine waters are compared on the basis of habitat volume, then fresh waters contain a concentration of fish species more than 5,000 times greater than that in the oceans. Clearly, the small portion of Earth that is covered with fresh water plays a disproportionate role in supporting biodiversity.

Michael McCloskey (1995) judged wild rivers to be the benchmarks of the healthy freshwater habitats that best support wild species. As part of the World Wilderness Congresses, he initiated an effort to identify the world's remaining wild rivers by examining maps for evidence of disturbance to banks, streambeds, and natural flow regimes (McCloskey, 1995, 199x). The procedure aims to identify natural reaches of streams as short as 50 kilometers, and is basic to the process of identifying targets for conservation.

In order to extend the analysis to include lakes and wetlands as well as streams, a complementary effort was undertaken for the 8th World Wilderness Congress to identify large areas of the world where surface waters are most likely to remain in natural condition. Recent evaluations of terrestrial wilderness used global datasets of human impacts (e.g., population density, intact vegetation) to identify areas that have been modified, and the remaining area was interpreted to hold the last remnants of wilderness (Sanderson, et al., 2002; Mittermeier, et al., 2003).

In a similar manner, this study identified large freshwater wilderness areas by eliminating watersheds that have suffered moderate to high levels of human activity. The primary units of analysis were the small-scale watersheds from *HYDRO 1K* (2003). These units were overlaid with binary data on human population density (*LandScan 2002 Global Population Database*, 2003), and watersheds with more than three people per square kilometer were eliminated. Additional watersheds were eliminated if they contained more than 3% agriculturally modified land (data from *Global Land Cover Map for the Year 2000*, 2003). Large-scale catchments were removed from consideration as freshwater wilderness if they were rated as moderately or strongly impacted by channel fragmentation or flow regulation (Nilsson, et al., 2005).

This study found significant areas of freshwater wilderness only at the northern extremities of the continents and in a few small watersheds of South America (see Figure 1). The loss of wild water in much of the tropics is associated with the fact that rivers have often served as avenues

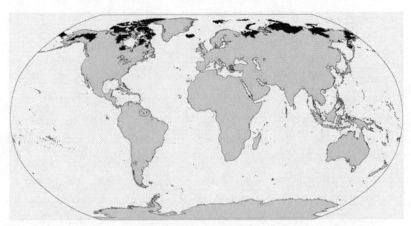

Figure 1 Large-scale areas where freshwater habitats are likely to remain in wilderness condition. Map by Erica Ashkenazi, GIS and Mapping laboratory, Conservation International

for human penetration of tropical forests (Sanderson, et al., 2002). Although Amazonia and the Congo Basin are considered terrestrial wildernesses (Mittermeier, et al., 2003), they are impacted by river channel fragmentation and human management of flow regimes. These effects are propagated over large distances within catchments. Southeast Asia, which has some of the world's highest levels of freshwater biodiversity, has long been impacted by agriculture and high human population density. It is now subject to high regulation of water flows at all scales.

In the northern hemisphere, population density and agriculture eliminate wild waters at mid latitudes. Regulation of water flow extends human impacts to even higher latitudes, leaving only a few northward-flowing watersheds with large-scale wilderness conditions. This study shows that wild fresh water is the rarest form of wilderness and that it is the form that is disappearing most rapidly. We have only a few years in which to protect the remaining wild water.

References

Global Land Cover Map for the Year 2000. 2003. GLC 2000. Database, European Commission Joint Research Centre.

HYDRO 1K Elevation Derivative Database. [downloaded 2003] Sioux Falls, SD. EROS Data Center. <http://edcdaac.usgs.gov/gtopo30/hydro>.

LandScan 2002 Global Population Database. 2003. Oak Ridge, TN: Oak Ridge National Laboratory.

McCloskey M. 1995. Wild rivers of the North: A reconnaissance-level inventory. In Arctic Wilderness, Martin V.G., Tyler N., editors, 130–138. Golden, CO: North American Press.

McCloskey M. 199x.

Mittermeier R.A., Mittermeier C.G., Brooks T.M., Pilgrim J.D., Konstant W.R., da Fonseca G.A.B. & Kormos C. 2003. Wilderness and biodiversity conservation. Proceedings of the National Academy of Science 100(18):10309–10313.

Nilsson C., Reidy C.A., Denisius M. & Ravenga C. 2005. Fragmentation and flow regulation of the world's large river systems. Science 308(5720):405–408.

Sanderson E.W., Jaiteh M., Levy M.A., Levy K.H., Wannebo A.V. & Woolmer G. 2002. The human footprint and the last of the wild. Bioscience 52(10):891–904.

Oceans as Wilderness
A Global Overview

Bradley W. Barr

Senior Policy Advisor,
NOAA National Marine Sanctuary Program

For nearly fifty years, the topic of ocean wilderness has been on the table. Since Wallis (1958) encouraged the National Park Service to include underwater areas in their thinking about potential wilderness, ocean wilderness has bubbled to the surface periodically, particularly during the World Wilderness Congresses. There is a limited but robust literature on the topic, and a handful of folks continue to keep the conversation going by making contributions to that body of work. There was a flurry of activity from 2000 to 2001 within the ocean conservation community in the United States, developing new initiatives around ocean wilderness, creating new organizations and coalitions devoted to advancing the idea, and initiating public awareness campaigns. But now, five years later, little remains of this flash of activity. Ocean wilderness is an idea that has not attracted sustained interest within the agencies that implement wilderness designations or within the larger wilderness community; nor has it ignited the fires of public interest. One could conclude that the lack of traction for the ocean wilderness concept is a result of it not being needed, wanted, nor valued. However, every time the idea re-emerges, it is met with enthusiasm, even passion. More analysis is done, more critical and creative thinking is applied to the topic, a few more issues addressed, and small steps forward are taken. The seeds of some great ideas simply take longer than others to germinate.

What Is Ocean Wilderness?

"At one extreme, wilderness can be defined in a narrow legal perspective as an area possessing qualities outlined in Section 2(c) of the Wilderness Act of 1964. At the other extreme, it is whatever people think it is, potentially the entire universe, the *terra incognita* of people's mind (Hendee and Dawson, 2002)."

One issue at the core of the punctuated history of interest in ocean wilderness is likely to be the lack of a broadly accepted definition. Defining wilderness is no trivial task. Wilderness is as much a perception as a place. It is inextricably tied to humans and the way we view the world around us. Any useful definition will therefore be predicated by who we are, our traditions and cultural heritage, and our connection to any place we might call wilderness. These are not simply places that are self-willed, where the direct influence of humans is limited and ecological processes occur naturally, although this is one of the core values of wilderness, but places we value for other reasons as well. We value wilderness for the awe it inspires, the humility it brings to us, the solitude it provides, and the sense of exhilaration from the real danger it offers. These values come to us both when we are fortunate enough to experience wilderness ourselves, but also when we step away from our daily lives and remember past wilderness experiences, or dream about what we believe being in the wilderness would be like even if we have never been there ourselves. Wilderness, however we perceive it, is part of a shared cultural history and tradition, and many people have a deep connection to these areas. It is the spiritual and cultural connections that make wilderness more than a place, and this is why defining what we mean by wilderness is very challenging. Therefore defining ocean, or any wilderness without incorporating this connection to the human spirit, to why we value wilderness misses the point entirely.

However, we humans are a polyglot of cultural diversity and perspectives. While any one person might have a very clear and unambiguous idea of what constitutes ocean wilderness, a room full of people are likely to have as many definitions or hold the opinion that wilderness is not a concept that is water soluble. The challenge is to construct a robust definition and foster consensus around that definition, first informally within the wilderness community, then through the public policy process and to memorialize that definition in statute, regulation, or agency policy. It is somewhat surprising to note that for all periodic, albeit brief, occasional enthusiasm about ocean wilderness, few definitions have been formally proposed and to this point in time, nowhere in the world has any consensus on a definition been reached.

Notwithstanding this absence of a consensus definition, a number of places that include ocean and coastal waters have been identified as ocean or marine wilderness, and a few areas are actually designated under law as wilderness, although generally associated with adjacent lands or islands that are also identified as wilderness. Two places that self identify as wilderness are the Tortugas Ecological Reserve, part of the Florida Keys National Marine Sanctuary, and the Northwestern Hawaiian Islands Coral Reef Ecosystem Reserve. Both of these areas are described as wild, remote, undisturbed, and rarely visited, words that are routinely used to describe wilderness on land. These descriptions convey some of the sense of awe and wonder surrounding these areas. Other areas have been similarly identified (Barr, 2001), and such information can be instructive in formulating a more formal and robust definition.

Ocean Wilderness around the World

Globally, there is but a single example of a site designated as ocean wilderness. Clifton (2003) mentions the designation and management of a number of marine wilderness zones within Indonesia's Wakitobi Marine National Park, established by Indonesia's National Act No. 5/1990, Conservation of Living Resources and Their Ecosystem. Clifton reports, however, that, "Despite the wilderness and rehabilitation zones being theoretically out of bounds, local fisherman still use these on a daily basis." Stakeholder involvement and awareness of rules regarding the use of marine wilderness zones is an issue, as is enforcement. Little published information is available regarding these areas, beyond the work of Clifton, but there are likely more than a few lessons learned from the Indonesian experience.

There are only a dozen or so countries that currently have wilderness legislation, and while all of those laws are focused on preserving terrestrial wilderness, nearly all of those countries have coasts and oceans. Many terrestrial wilderness areas are in the coastal zone, and more than a few of those areas have boundaries that end at the shoreline. The ecosystem linkages between land and water are well known, and it would be very surprising if the thought that wilderness should extend into the water was not becoming more broadly evaluated and discussed. Mention was made at the World Wilderness Congress by Murphy Morobe, Director of Communications for the President of South Africa, of a possible marine wilderness designation in South Africa. More proposals will surely follow.

Wilderness Waters in the U.S.

A small number of areas in coastal and ocean waters are designated as wilderness in the U.S. Table 1 provides information on areas designated wilderness waters under the U.S. Wilderness Act of 1964.

While this inventory represents a work in progress, as this inventory was based on available information provided by the U.S. Fish and Wildlife Service and National Park System, and may not represent all the sites so designated. The inventory does indicate the extent of existing wilderness waters designations in the U.S. and offers some descriptive information about those areas.

While it is difficult to fully reconstruct the record for these designations, it appears that only a few of these water areas were intentionally included by Congress when the designation was made. And, where they were intended, little in the way of specific direction was provided regarding any special management beyond the general guidelines provided by the Wilderness Act and agency policy developed under that law. Work is currently underway to clarify any conflicts in the data and better identify how these areas are managed. What can be inferred from this with regard to defining ocean wilderness is simply that it is possible to designate coastal and ocean waters as wilderness under the U.S. Wilderness Act, but the act offers little in the way of clarity regarding how wilderness values should be preserved in these areas.

Table 1 U.S. Wilderness Waters Inventory

Agency	Site	State	Total Acres	Wilderness Waters (a.)	Notes
FWS	Chassahowitzka NWR	FL	30,843	Yes (c. 11,600)	Waters of the GOM in Homosassa Bay
FWS	J.N. "Ding" Darling NWR	FL	6,391	Yes (2,825)	Embayments, mangroves, and tidal creeks
FWS	Blackbeard Island NWR	GA	5,618	Yes	Marshes and tidal creeks
FWS	Wolf Island NWR	GA	5,126	Yes	Marshes and tidal creeks
FWS	Cape Romain NWR	SC	65,269	Yes (29,000)?	Site staff claim no waters designated, file info conflicting
FWS	Monomoy NWR	MA	2,702	Yes (1,000+)	Wilderness boundary appears to be MLW ... some marshes and Hospital Pond "1000+" acres questionable
FWS	Alaska Maritime NWR	AK	3,465,027	Yes (264,512)	Simeonof Island 16,749 a., Semidi Island 244,656 a.
NPS	Everglades NP	FL	1,508,537	Yes (625,000)	Wilderness submerged lands but not surface waters
NPS	Lake Clark NP&P	AK	4,030,025	Yes	Upper reaches of Tuxedni Bay ... State of AK objected to inclusion of any tidelands in comment letter regarding wilderness designation
NPS	Pt. Reyes NS— Philip Burton Wilderness	CA	71,068	Yes (8,233)	43 USC 1340(h) prohibits oil and gas leasing within 15 mi. of wilderness, unless State of CA permits
NPS	Glacier Bay NP&P	AK	3,244,840	Yes (53.270)	Wilderness waters subject to extensive special management measures

A Working Definition

As a prelude to the 8th World Wilderness Congress, The WILD Foundation sponsored an International Wilderness Law and Policy Roundtable in Washington, D.C. (November, 2004), and a theme session was held as a part of the roundtable to discuss defining ocean wilderness. The marine theme panel convened international experts from the U.S., Canada, Australia, and Chile, in the field of marine protected areas science and management, and those who had published scholarly works related to the topic of ocean wilderness.[1] This group addressed the following questions:

- Can we develop, and if so what might be a working definition for ocean wilderness?

- What are the relevant wilderness values that might be most important in preserving ocean wilderness?

- What are other challenges and unresolved issues related to effectively addressing the designation and preservation of ocean wilderness?

Over three days, these questions were discussed and debated, and presentations were made on relevant research, policy development, and management activities. Given the scope of the questions before them, participants made significant progress and reached consensus on a number of key points:

- Regarding existing authorities, U.S. Wilderness Act, where jurisdiction exists (e.g., national parks, national wildlife refuges), is indeed relevant and can be directly applied to management of ocean wilderness. This conclusion was interesting in that it was in direct contradiction to the consensus reached by a previous panel of non-governmental organizations meeting to discuss the concept of ocean wilderness, as reported at the Roundtable (personal communication, David Festa, National Marine Program Director, Environmental Defense, November 8, 2004).

- In developing definitions, it is essential to consider that ocean wilderness is multidimensional, involving consideration of areas above, on, and below the sea surface, and perhaps requiring differing management approaches.

- Outcomes are more important than words. Given the political sensitivity of wilderness and the divisive process that is likely to accompany expanding Wilderness Act authorities, as well as the potential adverse impact to other wilderness programs during such deliberations where Wilderness Act jurisdiction does not currently exist, a better strategy might be to apply wilderness values through zoning, without evoking the word wilderness in that process. This is done quite successfully in the rezoning initiative at the Great Barrier Reef Marine Park (personal communication, Jon Day, Director of Conservation, Biodiversity and World Heritage, Great Barrier Reef Marine Park Authority, November 9, 2004).

The product of the deliberations of this panel was a working definition of ocean wilderness, linked to the wilderness values (Table 2) the definition was intended to preserve. The definition is as follows:

"Areas of the marine environment that are untrammeled and generally undisturbed by human activities and dedicated to the preservation of ecological integrity, biological diversity, and environmental health."

An area of ocean wilderness may provide:

1. Opportunities for quiet appreciation and enjoyment in such a manner that will leave these areas unimpaired for future generations as marine wilderness

2. Continued opportunities for subsistence uses and indigenous cultural practices

The panel felt that a number of issues would require further discussion:

- Recreational Fishing: The panel concluded that recreational fishing should not be permitted simply because of political

Table 2 Wilderness Values Underlying Roundtable Working Definition

Key Element of Definition	Wilderness Value	Relevant Descriptive Terms
Untrammeled and generally undisturbed by human activities	Recreational Scenic Recreational Therapeutic	Challenge, risk, adventure, compatible hunting and fishing Remoteness, enjoyment without motorized water sports
Preservation of ecological Integrity	Ecological Scientific Educational	Intact flora and fauna, store house of biodiversity, allow for restoration, maintain unimpeded ecological community interactions Research and monitoring Creating understanding and respect
Opportunities for quiet appreciation and enjoyment	Spiritual Artistic	Well-being, healing, inspiration Inspiration, imagination
Unimpaired for future generations as ocean wilderness	Existence Moral Historic Symbolic Economic Ecological Spiritual	Peace values Allowing appropriate recreational activities Well-being, healing, inspiration
Opportunities for subsistence Indigenous cultural practices	Cultural Aboriginal rights	In perpetuity, perpetuating myth and legend Uphold treaty rights, laws, subsistence uses

pressure. It can be effectively managed with gear, time, area access restrictions, and other traditional recreational fishery management measures. But how much active management should be undertaken in ocean wilderness?

- Motorized Access: Motorized access is probably necessary for safety and practicality in most ocean areas, but where vessel size, draft, power, and the like can be limited it should. Non-motorized areas should be considered where appropriate, as has been done in certain wilderness waters of Glacier Bay National Park and Preserve.

- Determining How Disturbed an Area Could Be and Still Warrant Wilderness Designation: Determining how much, how long, how frequent and how extensive human disturbance would have to be in an area, and how long would it take to recover to a natural state, would be a prerequisite to making a decision as to whether an area could be designated ocean wilderness. Having some sense of disturbance history and recovery rates would be essential, but may not be readily available in the majority of cases. Just seeing the extent of human disturbance might be difficult, as these areas are largely underwater and may require advanced undersea technology to characterize and map resources.

The panel made a good first effort at tackling this difficult task of defining ocean wilderness. It was able to come to full consensus on a working definition, and began to identify the wilderness values significant to ocean wilderness; this was a significant accomplishment. It is clear from the work of the roundtable theme panel that the task of defining ocean wilderness does not have to start with first principles. Important lessons can be learned from the long and often difficult work of terrestrial wilderness scholars and managers, lessons that can be directly applied to the task at hand. Those involved in protecting and managing ocean ecosystems have already begun to delve into this issue of applying the wilderness concept to ocean waters and some small steps are being taken to identify key questions. Ever pragmatic, marine protected area managers are already preserving ocean wilderness in places like Glacier Bay and the Great Barrier Reef, unwilling to wait for the theory to catch up to the pressing need. Different means to the same end.

What we call something is important. Oceans are public waters, in common ownership, and the public has an important stake in these deliberations. Provocative to some, evoking the term wilderness in this work is both necessary and appropriate. The public, as stewards and owners, must be fully aware of this task and its implications so that they can effectively engage in the public policy process as informed and responsible participants. There should be little trepidation about the use of the word "wilderness," as the public has a deep and visceral connection to wilderness heritage. If anything, it should help to insure that a broader spectrum of interested public will be actively engaged.

Looking to the Future

Arriving at some consensus on a definition of ocean wilderness is only the first step in a long journey. While we are sometimes impatient at the pace at which any type of wilderness is designated, it is good to be reminded that the U.S. Wilderness Act was only passed by Congress in 1964, and most wilderness legislation around the world is more recently enacted. In the U.S., marine protected areas were identified in national parks and national wildlife refuges for only about a decade longer than the Wilderness Act, and the U.S. National Marine Sanctuary Program was established in 1972. Good ideas sometimes are slow to take hold, and even slower to grow into thriving programs.

For now, the work on defining ocean wilderness is progressing, and consensus seems like a real possibility. More work needs to be done to study and evaluate existing wilderness management for those coastal and ocean areas already designated under law. More thought needs to be given to what is the ultimate objective to designating ocean wilderness. There are truly awe-inspiring diverse and productive areas in the oceans of the world profoundly important for maintaining ecosystem structure and function and the services they provide. These areas are, without question, the equals of our most valued terrestrial protected areas. Coral reefs, deep-sea hydrothermal vent ecosystems, offshore banks and upwelling areas, seamounts and canyons are among potential candidates. But which of these, if any, are valued for their wilderness qualities? Which are best preserved as wilderness, or are other resource management strategies more appropriate? Even if we come to consensus on a definition, can potential ocean wilderness areas find their way through the political process of designation, especially at a time when proposals for selling off national parks are being tabled in the United States Congress? Perhaps it is well and good to work purposefully for now, and hope that the climate will become more favorable in the future. There is always the risk of losing the opportunity to preserve these ocean wilderness areas as a result of the continued degradation of ocean resources, overexploitation, and misuse. However, what sets wilderness apart is that it gives us hope. Hope that our children and their children will be able to experience the wilderness we have, hope that our deep and visceral connection to wilderness will lead us to embrace preservation over exploitation, and the hope that there are still, and will continue to be, blank spots on the map … both on the land and in the water.

~

The views expressed herein are those of the author and do not necessarily reflect the policies, positions, or views of the Department of Commerce, NOAA, or any of its sub-agencies.

References

Barr B.W. 2001. Getting the job done: Protecting marine wilderness. Pp. 233–238 in Harmon David, editor, Crossing Boundaries in Park Management: Proceedings of the 11th Conference on Research and Resource Management in Parks and on Public Lands. The George Wright Society, Hancock, MI.

Clifton J. 2003. Prospects for co-management in Indonesia's marine protected areas. Marine Policy 27:389–395.

Hendee J.C. and Dawson C.P., editors. 2002. Wilderness management: Stewardship and protection of resources and values (3rd ed.). Golden, CO: Fulcrum Publishing.

Wallis O.L. 1958. Research and interpretation of marine areas of the U.S. National Park Service. Proceedings of the Gulf and Caribbean Fisheries Institute 11:134–138.

Marine Wilderness
Protecting Our Oceans Is Protecting Our Land

David Rockefeller, Jr.
Board of Directors, National Park Foundation

Nobody washes a rental car. Nobody picks up random trash from the sidewalk. When property does not belong to us, we are unlikely to look after it.

This is the dilemma of "the commons," and especially of the oceans. Although international law for decades has decreed that national boundaries extend 200 miles into the contiguous ocean, only fishermen think of these waters as belonging to them. In fact, these waters belong to all of us, and we would do well to create an "ocean ownership ethic" in our peoples. Otherwise, who will care for them?

In my three years of service on the Pew Oceans Commission, I learned so much about public attitudes in the U.S. toward ocean life: that we take it for granted and can't really imagine that human activity, what we put in or what we take out, could have a serious impact on the great oceans. Everything we do has the potential to do great harm: a single spilled barrel of fuel or road chemicals or the sprays and fertilizers we apply to our crops, some of which run into the nearby rivers, thence to the oceans; or the corals and exotic reef fish we extract and sell to private aquarium owners, no less the tons of pelagic food fish scooped from the deep sea by omniscient and omnivorous factory trawlers.

A recent science journal article described how in five large zones of the open ocean the introduction of factory fishing fleets reduced the population of large target fish in only fifteen years by 90%! How disastrous and how shortsighted that the ocean protectors, there were fewer of them in the 1950s and 1960s, could have let this happen. "Never again," we ought to be saying.

The fact is that three twentieth-century forces have conspired to challenge the ocean's protein supply: global population growth from 2 to 6 billion in just 100 years, increased purchasing power in many parts of the hungry world, and highly sophisticated technology for finding, harvesting, processing, and preserving fish at sea. Many current practices are unsustainable, and something must be done.

One of the solutions put forward by the Pew Oceans Commission is the creation of ocean wilderness zones, what are frequently referred to as "marine reserves." Just as we fallow our fields to let them rejuvenate and just as we have created game reserves on land, we need to develop more regions in the ocean that are free from human predation. These should be created under both national and international jurisdictions, since great swatches of our oceans lie beyond the 200 mile "exclusive economic zones" of all nations.

Of course, there are many reasons why we terrestrial creatures should care about the health of our oceans. Just as we have discovered the rain forests to be rich repositories of new species of life, both fauna and flora, scientists estimate that as many as a million unclassified species inhabit the ocean deeps, until recently beyond our grasp (even beyond Sylvia Earle's). Compare that number to the fewer than one dozen new species thought to inhabit planet Mars, and we ought to ask ourselves, especially here in the U.S., whether we don't have our research priorities upside down. Should we not be investing more in the undiscovered depths of Earth than we are proposing visits to the other spinning sisters of our solar family?

Not only are there medicines to be found in the oceans' deeps, but new forms of life that may unlock the mysteries of how this globe was born, or how life can exist in extremes of heat, cold, darkness, and pressure. In the future, when we discover metaphoric "goldmines" of new minerals in the oceans' deeps, the international agencies must have in place stronger regimes that protect and regulate who can go there, who can extract, and with what recompense paid to the official protectors of our deep wilderness.

Alaska is well known as the northern counterpart to the California Gold Rush of the mid nineteenth century. Let us take a lesson from the sad scramble that ensued when those prospectors entered these unregulated mountains and rivers of hope. Alas, that scramble continues today here in Alaska, and still threatens to poison and pillage some of these wildlands and waters. Deep ocean minerals are part of "The Commons." They are not the province of looters.

In the past year, the great tsunami in the Indian Ocean and two coastal hurricanes in the Gulf of Mexico have reminded us all that the ocean's power can wreak havoc upon the land and especially its coastal residents. Meanwhile, the rapid melting of the Arctic ice cap and neighboring Greenland ice sheet, alert us to the growing global threat to lowland coastal inhabitants for whom even inches of sea level rise could produce mass homelessness and devastation.

The tsunamis have no proven origin in human behavior, although they should cause all lowlanders and their governments to think more about "preparedness." But, vicious hurricanes ("typhoons" in the Pacific) and rapid ice melt are increasingly seen by the scientific community to be products, in significant proportion, of human behavior that we do have the power to mitigate. The oceans are like a miraculous balance wheel in global meteorology, warming the cold places and cooling the warm, but they are themselves vulnerable to human inputs, especially those relating to energy exhalations. It is increasingly clear that the global community must make significant adjustments. Nowhere is that more true, of course, than in the biggest consumer nations.

So we must be mindful of our oceans, not only for the meals they bring us, but also for their medicines and minerals, as well as their meteorological surprises.

The availability of human wilderness experience is another issue of importance to me. I believe there are intrinsic benefits to be derived, not only by the wild creatures of land and sea but also by the human souls who take the risks and expend the energy to enter wilderness zones. Some of our greatest literature and music and, indeed, philosophy and religious revelation have been produced as a result of a solitary writer's wilderness exposure. "Deprivation, challenge, silence, contemplation, direct contact with the natural world"—these are the elements that seem to produce human creativity from the wilderness experience. They are priceless, and they must be cherished.

It seems that the media—the print media to some extent, but especially the film media—have recognized the importance of creating greater public appreciation for the interaction between the ocean wilds and the creatures that inhabit them, or who, like us humans, depend upon them.

Perhaps Peter Benchley's *Jaws* was the first to "break the surface," you might say, of human awareness, but recently we've been offered the animated *Finding Nemo*, the French films *Winged Migration* and *March of the Penguins*, and the BBC's *Deep Blue*.

Each of these films, in very different ways and voices, has contributed to popular consciousness a greater understanding of the deep oceans and how they represent complex ecosystems in which predators and prey, denizens of the air and land and sea, habitat and inhabitants, all interact in a miraculous and sometimes brutal fashion.

I am very grateful for those artists who have taken the trouble, and often the personal risk (waiting in shark cages or flying in ultra-light aircraft) to present these dramatic examples of life in and around the marine community.

I cannot claim to have experienced the ocean wilderness myself in anything like the scale Mr. Nainoa Thompson has accomplished. It is truly spectacular what he and his canoeing colleagues have managed to do in the ocean, while forsaking all the modern instruments of navigation. My sailor's hat is off to him!

And I add my salute to Sylvia Earle, the Jacqueline Cousteau of our time, and Clem Tillion, long-time fisherman in some of the world's most dangerous waters.

As a more traditional sailor, who chooses to remain on the sea surface and avoid the icy winter storms, I have been very fortunate to explore many of the world's briny places in person. Beginning in the 1950s I have ventured the translucent Caribbean Sea; the turbulent western Atlantic from Florida to Labrador and as far out to sea as the Island of Bermuda; 3,000 miles of the amazing Gulf of Alaska in a convoy of four small sailboats; the North Sea and eastern Atlantic surrounding the outer islands of Scotland; and the history rich central and eastern Mediterranean. I have sailed at night in storms, for days in dense fog, among icebergs, alongside pods of porpoise and whales; on tiny planning boats and on tall vintage schooners; alone and with dozens of fellow crewmen; as captain and as "rail meat."

And always when at sea, I have the feeling of immense good fortune that there is still an oceanic wilderness available to those who dare and those who come to learn in the cradle of the sea. As the world's human population grows, and as the places diminish where you can still find "silence and the night sky," it is essential for our individual and collective human psyches that these sacred wilderness zones—both on land and on sea—be preserved and protected.

It occurs to me, in fact, that the ocean wilderness is precious for two opposite reasons: for the precious knowledge it will reveal to us as we explore and protect it, but equally for the mystery it represents that we cannot even fathom. Being in love with the ocean is like a great love for a woman, you know. You want to learn more and more about your partner, but not everything. So please leave room for the mystery. Love your ocean. Explore and protect it. But always honor the mystery of it.

For all these reasons: economic, nutritional, biological, and psychological, I urge the imperative of the "preservation of marine wilderness." Our oceans belong at once to no one and to everyone. It is our collective responsibility to preserve and cherish the gift of the deep ocean. For those of you whose professional experience is in the preservation of wildlands, I thank you and I urge you to extend your radius of consideration to the wild oceans. For those of you who make your living from the sea, I also thank you and I urge you to join with other ocean advocates to protect the very golden eggs that lie vulnerably along the ocean floor. You fisher people, tourist tradesmen, innkeepers, boat builders, as well as you sailors and surfers, and others who find recreation at sea, all of you, all of us, must be the advocates of saving for the ages what we have been so fortunate to enjoy in our own lifetimes.

For my own part, I have founded sailorsforthesea.org, with this imperative in mind. On land, it has often been the forest hunters and stream fishers who have saved our wild places. Why should the same principal not apply to the oceans? Sailors for the Sea is a web-based education resource for sailors and other recreational boaters who want to learn about the challenges to their ocean highways, and then become actively involved locally, regionally, or nationally in addressing these challenges. As we say at Sailors for the Sea, "Sailors need the ocean and the ocean needs us."

In conclusion, I submit we are all literally "in the same boat," and when the blue water spills over the coaming, we all must bail together. The conservation movement globally has done a spectacular job of focusing the eyes of the world on the interaction of healthy land-based ecosystems and healthy human populations.

Now it is time for us, feet planted firmly on the ground, to reach our arms out to the oceans and express our gratitude for all the "meals, the medicines, and the mystery" they represent; to express our gratitude in the most meaningful way we can by vowing to protect them forever.

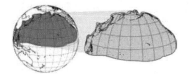

The Status of Pacific Salmon and
Their Ecosystems across the North Pacific

Xanthippe Augerot

Director of Conservation Programs, Wild Salmon Center

Salmon are the iconic species of the North Pacific. Like monarch butterflies or grey whales, salmon migrate thousands of miles across political boundaries and link the countries and economies of the North Pacific. Their cultural and economic importance cannot be underestimated; they provide jobs, food, recreation, and inspiration (Augerot, 2005). Pacific salmon also serve as an indicator of clean, cool water and healthy natural rivers. They may persist in river systems that have been ditched, diked, and paved, but they will be present in very small numbers and with truncated biodiversity, making them much more vulnerable to human or natural disturbance. Robust, abundant, and diverse salmon populations are symbolic of wild rivers with unaltered natural hydrological processes (Stanford, et al., 2005).

There are eight widely distributed species in the genus *Oncorhynchus*: pink, chum, sockeye, chinook, coho, masu, steelhead-rainbow, and cutthroat. They are distributed around the North Pacific from Taiwan to Mexico, though the small populations at the southernmost part of the range on either side of the Pacific are now landlocked, no longer going to sea. Several species also form small populations in the Arctic; populations that some believe are growing with warming Arctic waters. Pink and chum salmon are the most abundant and the most widely distributed species, while sockeye commands the highest market premium. (See Figure 1: Range map, all species).

Temperate river ecosystems co-evolved with Pacific salmon, from Korea to California. As many as 138 species of animals rely upon salmon-derived nutrients, carried from ocean rearing grounds to freshwater each year. Some of these relationships are direct, such as char predation on salmon eggs during spawning; others are indirect, such as the spiders that depend on lush streamside willows, amply fertilized by salmon carcasses (Cederholm, et al., 2000). The spiders in turn serve as prey for fish and songbirds. Although we are just beginning to accumulate evidence from intact salmon river systems about the nutrient-pumping role played by salmon (Gende, et al., 2002), we speculate that many of our salmon-impoverished systems, such as those in the Oregon Coast range, may be less productive now than historically in part due to the long-term depression in salmon-delivered marine nutrients (Gresh, et al., 2000).

Pacific salmon evolved in the cold waters of the North Pacific over millennia, enduring dramatic environmental changes in sea level and climate, persisting due to their adaptability. Their resilience is embodied in their genetic diversity at the species and population level, and their varied life history strategies. Chinook salmon may spend one summer or one and a half years in freshwater, and from three to five years at sea. Some return in the spring, others in the summer, fall, or even winter

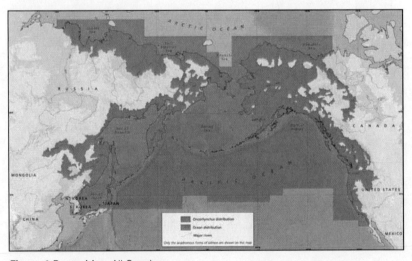

Figure 1 Range Map, All Species

(i.e., Sacramento River, California; Healey, 1991). This diversity of life histories spreads risk of extinction for any given Pacific salmon species over both space and time. It also ensures the delivery of marine-derived nutrients to many rivers, especially large temperate systems such as the Sacramento River (U.S.), the Columbia River (U.S.), and the Fraser River (Canada), virtually year-round. More northerly systems, including major producers such as the Kuskokwim (U.S.), the Anadyr (Russia), and the Amur (Russia and Peoples' Republic of China), experience more compressed but very abundant salmon runs from mid summer tailing off through the fall.

While local adaptations are a blessing for the resilience of the genus *Oncorhynchus*, they also make it very difficult to select the most appropriate conservation unit for the determination of status and trends. The IUCN World Conservation Union Red List Categories and Criteria provide a global system for classifying species risk of extinction (IUCN, 2001). However, determination of extinction risk at a global level for any given salmonid species has very little meaning for Pacific salmon species. For example, by the time coho salmon could be judged to be vulnerable to extinction globally on the basis of population reduction or drastic reduction in their extent of occurrence, many locally adapted populations are likely to be critically endangered or extinct in the wild. Subpopulations at the genetic resolution of one migrant per year are difficult to determine given present salmonid genomics and technology. Therefore, our first attempt to assess Pacific salmon status throughout their geographic range was largely qualitative.

Previous Analyses

In 1991, Nehlsen and her colleagues declared that Pacific salmon (*Oncorhynchus spp.*) in California, Oregon, Idaho, and Washington had reached a crossroads where immediate recovery options were still available but soon would be lost if habitat loss and human population growth continued apace. This first-ever regional synthesis of local biological information pointed to a widespread pattern of decline of more than 200 stocks at risk of extinction or of special concern. Other reviews have since identified additional at-risk stocks both in the United States and Canada (Baker, et al., 1996; Slaney, et al., 1996). Reviews by the U.S. National Marine Fisheries Service (www. nwr. noaa.gov/1salmon/ salmesa/pubs. htm) confirmed the general trends described in 1991 and have led to the formal listing of twenty-six evolutionarily significant units of Pacific salmon under the U.S. Endangered Species Act. In British Columbia and the Yukon, Slaney, et al. (1996) classified nearly 17% of more than 5,000 salmon stocks as at some risk of extinction or concern. Baker, et al. (1996) concluded that among the spawning aggregates of southeast Alaska, only about 1% are at a moderate or high extinction risk. These regional assessments suggest a broad gradient of increasing risk of extinction in North America from southeast Alaska to California.

The most comprehensive Russian summary of stock status is based upon analysis of catch trends across the Russian Far East (Radchenko, 1998). Radchenko attributed most of the variability in Russian salmon catch to climate effects. As of 1996, pink (*Oncorhynchus gorbuscha*) and sockeye (*O. nerka*) stocks were assessed to be stable or increasing, but chinook (*O. tshawytscha*) and coho (*O. kisutch*) were declining in abundance. Chum (*O. keta*) salmon stocks were also in decline, particularly in the southern Russian Far East. Masu (*O. masou*) trends were not assessed.

Unfortunately, trends in salmon abundance (catch plus reproducing adult spawners) are unknown or unreported for many river basins around the North Pacific. Commercial catch trends have been documented (e.g., INPFC, 1979; Fredin, 1980; Shepard, et al., 1985a), but the species composition and basin of origin of this catch is often poorly known (Shepard, et al., 1985b). Catch is usually aggregated by statistical districts conforming with political jurisdictions, rather than the distributions of naturally reproducing populations or genetically distinct stocks.

If status is viewed through the lens of commercial catch, the overall picture is fairly optimistic. Since 1920, despite significant variability, total catch rates have continued to climb (Table 1). However, the current high levels of productivity are driven largely by two species, pink and chum salmon, which are the most abundant species and form large

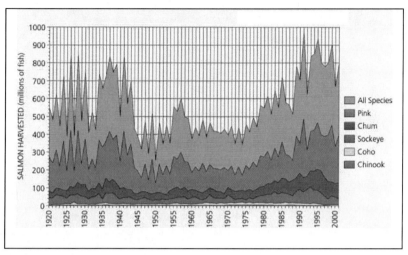

Table 1 North Pacific Salmon Harvest, 1920–2001

aggregations at northerly latitudes. Sockeye salmon has also been very abundant in the past two decades. Coho and chinook catches are both in decline (NPAFC). Catch records for steelhead, primarily a sport fish species since the 1920s, and for masu salmon are poor, in part due to their low natural abundance.

Catch statistics as a metric for status or trends can be very misleading. Catch statistics are most reliable for sockeye salmon, traditionally the most highly valued salmon species, with the most effort deployed in stock forecasts for the fishing industry. They best represent areas that produce large volumes of salmon, and poorly represent low and moderate abundance regions, which often register the first symptoms of decline when climate conditions are poor.

Catch statistics also count artificially cultured hatchery fish alongside wild naturally spawning salmon. Figure 2 depicts the distribution of hatcheries around the North Pacific. Hatcheries proliferated at a rapid rate in the 1960s and 1970s, as hatchery technology improved while global catches were depressed, due to poor ocean rearing conditions. Hatcheries represent a wide-range political response to contending with large commercial fishing fleets and declining salmon abundance (Augerot, 2000).

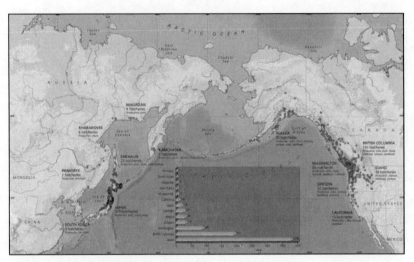

Figure 2 Salmon Hatcheries

Distinguishing wild from hatchery fish is not a simple task, as hatchery fish are not routinely marked and tracked everywhere around the North Pacific. On the basis of 1977 to 2004 catch data and information about hatchery return rates, Ruggerone and Goodman (2006) estimated that 80% of the approximately 506 million standing stock of North Pacific salmon are wild. However, the proportions vary dramatically by species and region. For example, Russian chinook salmon are assumed to be 99% wild, whereas U.S. Pacific Northwest chinook are only 57% wild, with almost half of the annual catch produced by artificial culture operations. The vast majority of chum salmon caught and reproducing each year are assumed to be of hatchery origin—63% of the total around the North Pacific.

This article reports results of the first attempt to assess the status of wild salmon stocks across their entire North Pacific distributional range, conducted between 1998 and 2001. Our intent was to determine whether coherent patterns of extinction risk emerge at this broad scale and whether any common causal mechanisms might account for salmon declines.

Methods

Salmon Ecoregions

A prerequisite for any trans-Pacific salmon analysis is to establish a consistent geographic framework and spatial resolution for integrating widely divergent sources of information. To minimize the effects of inconsistent methods and scales of resolution for characterizing a stock, we established a common geographic template for aggregating available data and assessing the status of salmon across their entire North Pacific range. In constructing this template, we made the assumption that salmon survival and recruitment are influenced significantly by freshwater and ocean conditions experienced throughout early juvenile stages (Karpenko, 1998; Pearcy, 1992). We used several existing classifications of marine circulation systems and production domains (e.g., Ware and McFarlane, 1989; Sherman and Duda, 1999) in a four-stage hierarchical classification to divide the North Pacific Rim into a series of salmon ecoregions, defined as watershed-coastal ecosystems of distinct physical characteristics, including the full sequence of riverine, estuarine, and near-shore marine habitats used by juvenile anadromous salmonids. The final criteria and boundaries for the ecoregions were developed in 1999 during a workshop of Japanese, Russian, Canadian, and U.S. scientists held in Corvallis, Oregon ("Workshop Participants;" see North Pacific Salmon Workshop Working Group, acknowledgments). Workshop Participants endorsed a four-level hierarchical classification defining salmon ecoregions:

- Level one includes two ecoregions: the Arctic Ocean and the Pacific Ocean, and associated freshwater drainages.

- Level two includes eighteen ecoregions, including semi-enclosed seas and primary ocean circulation systems with distinct processes or bathymetric characteristics in the North Pacific (similar to the Large Marine Ecosystems defined by Sherman and Duda, 1999), and associated freshwater drainages. There are two Arctic Ocean and sixteen Pacific Ocean regions defined at this level.

- Level three includes thirty-nine ecoregions, including finer-scale coastal discontinuities within each semi-enclosed sea or major circulation system such as fjords, straits, and areas with distinct production processes (e.g., upwelling and downwelling areas). There are three Arctic Ocean and thirty-six Pacific Ocean regions defined at this level.

- Level four includes sixty-seven ecoregions, including major drainage basin networks equal to or greater than the area of the Kanchalan River (22,230 km^2) entering each Level 3 coastal area. There are fourteen Arctic Ocean and fifty-three Pacific Ocean ecoregions defined at this level.

Although many ecoregions encompass large river basins draining distinctly different sub-basin ecosystems, no single trans-Pacific classification exists that relates sub-basin ecosystem variations to anadromous salmon stock characteristics. Therefore we arbitrarily defined Level 4 ecoregions as a convenient minimum river basin standard (i.e., Kanchalan River) for aggregating existing stock data. Basin area smaller than the Kanchalan River greatly increases the number of ecoregions and the complexity of any trans-Pacific analysis. There were twelve exceptions to this rule, areas delineating aggregations of small coastal drainages sharing near-shore conditions distinct from adjacent ecoregions.

We used the Environmental Systems Research Institute's (ESRI) Digital Chart of the World (DCW) to provide a digital geographic representation of the North Pacific, to identify the stream networks and spatial boundaries of each ecoregion, and to establish a geographically explicit data management system for our stock status information. The DCW scale of resolution includes streams to approximately fourth order (Strahler, 1957) on a 1:100,000 map. To define the geographic boundaries for each ecoregion, we drew digital polygons around the stream networks associated with each of the four levels of the hierarchy. To meet page-size limitations, our maps (Figures 1, 2, and 3) are shown in Mercator projection, which exaggerates the spatial extent of the northernmost ecoregions and diminishes the relative size of the Pacific Ocean with regard to continental land mass. (See Acknowledgments for workshop participants.)

Stock Definitions and Status Criteria

We instructed workshop participants who provided status information to use Ricker's (1972) commonly accepted definition that describes a salmon stock as a population that spawns in a particular river and season and does not interbreed with other spawning groups. Past salmon status reviews (e.g., Baker, et al., 1996; Slaney, et al., 1996), however, have acknowledged that there can be considerable variation in the number of stocks that are delineated depending on the resolution for which adult trends are monitored, life history or genetic distinctions are known. While highly generalized stock distinctions underestimate the diversity of populations and life-history types within a region, we believe the effects of variable data quality and stock specificity among regions around the North Pacific Rim are minimized by analyzing results based on the total proportion of all identified stocks within each ecoregion and risk category.

We evaluated stock status for seven anadromous species of the North Pacific: chinook, chum, coho, masu, pink, sockeye, and steelhead. Although our original survey requested assessments by seasonal race, we aggregated our results to the species level because the level of knowledge of status by seasonal race was uneven at best. Our analysis integrates data from previously published sources for river basins in the Pacific Northwest (Nehlsen, et al., 1991; Baker, et al., 1996; Slaney, et al., 1996) and includes new information we collected from management and academic biologists from Korea to California. All risk rankings are based upon the best expert judgment of local biologists, except for the rankings from British Columbia and Southeast Alaska, which use quantitative escapement data. The major data sources and approaches we used to characterize salmon status for all North Pacific regions and ecoregions are summarized in Table 2.

Although all jurisdictions responded to our requests for information, we were unable to assess stock status for every North Pacific ecoregion. Among the Pacific ecoregions, lack of trend data for wild stocks prevented us from categorizing salmon status in Korea and Japan. Very little is known about the status of salmon populations in remote Arctic Ocean drainages. Information for central and western Alaska was

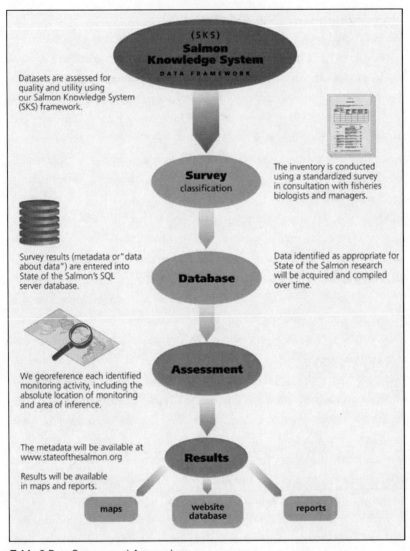

Table 2 Data Sources and Approaches

not accessible (and therefore not included in our analysis). Southeast Alaska stock status (Baker, et al., 1996) is reported by management areas, which could not be directly resolved to the boundaries of our ecoregions. Therefore, we report stock status for the region as a whole, combining portions of four ecoregions: Alaska Coastal Downwelling,

Transboundary Fjords, Nass-Skeena Estuary, and Nass River. Data for the British Columbia portions of the Nass-Skeena and Nass River ecoregions are reported separately.

We classified extinction risk for Pacific salmon into four categories adapted from Nehlsen, et al. (1991):

- Unthreatened: The stock is at no risk of extinction. It is abundant relative to current habitat capacity and not exhibiting any recent declines. It has never been previously identified at risk of extinction.

- Moderate risk: The stock is at moderate risk of extinction. After declining, the number of spawners appears to be stable and above 200 adults. Approximately one adult per spawner is returning to spawn.

- High risk: The stock is at high risk of extinction. The trend in the number of spawners is declining and less than one adult fish returns for each parent. Spawning populations of less than 200 adults within the past one to five years, in the absence of evidence that they were historically small, were given this rank.

- Extinct: The stock no longer reproduces in its historical range.

Threats to Salmon
During the North Pacific Salmon Workshop (May 4–6, 1999), we asked workshop participants to identify by ecoregion and within their respective regions of expertise the major threats to Pacific salmon stocks. We reorganized threats into nine major categories: agriculture, dams, hatcheries, logging, mining, oil and gas, over harvest, poaching, and urbanization. We excluded threats of climate change because climate fluctuations potentially affect all North Pacific stocks, though unevenly across space and time. We then ranked primary, secondary, and tertiary threats. Since more than one factor may contribute equally to salmon risk in an area, we did not limit the number of factors that could be designated as primary, secondary, or tertiary.

Results

Stock Status

We had sufficient information to assess the status of 7,801 stocks from forty-one of the fifty-three Pacific Ocean ecoregions (Table 2). Of these, we classified the stocks as:

- 282 (3.6%) extinct
- 883 (11.3%) at high risk
- 822 (10.5%) at moderate risk
- 5,814 (74.5%) at low risk of extinction

The ecoregions with the highest proportions of extinct stocks are concentrated in the U.S. Pacific Northwest (Figure 2). More ecoregions in Asia than in North America are classified entirely in the low-risk category. A group of ecoregions with a high proportion of moderate-risk stocks is centered around the Sea of Okhotsk (ecoregions 16–21) and in the western Bering Sea (ecoregions 23–24). Overall findings showed that:

- Pink are least threatened, with 93% of stocks classified at low risk of extinction
- Masu, with 76% of stocks, classified at low risk of extinction
- Chum, with 69%, classified at low risk of extinction
- Sockeye and coho, with 67%, classified at low risk of extinction
- Chinook, with 56%, classified at low risk of extinction
- Steelhead, with 54%, classified at low risk of extinction

The geographic patterns of stock status are described as follows by species.

Chinook

Chinook stocks are at greatest risk of extinction at the southern edge of their range on both sides of the Pacific, particularly in North America

(Figure 3). There are nine ecoregions with at least one extinct chinook stock, and four ecoregions where 100% of chinook stocks are categorized at low risk. All chinook stocks are extinct in the Weak Upwelling Cline (52) and the California Undercurrent (53). We identified no threatened chinook stocks in Chukotka (25–27) or in the British Columbia portion of the Nass River (40). Chinook in the Western Kamchatka Current (21) are at much higher risk than in surrounding ecoregions.

Chum

Chum stocks are at greatest risk of extinction at the southern edges of their range, particularly in North America. We classify seven ecoregions with at least one extinct chum stock and ten ecoregions at low extinction risk for all chum stocks. We did not identify any ecoregions where all chum stocks are extirpated. Notably, two chum stocks (11%) in Primorye (Liman Current, 7) and eight stocks in the Amur River (14) are considered extirpated, the only salmon extirpations noted anywhere in the Russian Far East. In a cluster of ecoregions around the Sea of Okhotsk (17–21) and in the western Bering Sea (23–24), we classified all chum stocks at moderate extinction risk.

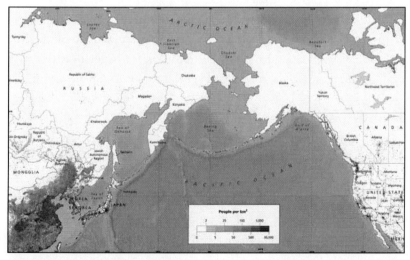

Figure 3 Human Population Density

Coho

Coho are at greatest risk of extinction at the southern edge of their range in North America. In eight ecoregions, the extinction risk is low for all coho stocks. All coho are extinct in the Weak Upwelling Cline in California (52), and at least one coho stock is extinct in eight additional ecoregions. The Columbia River and Sacramento-San Joaquin ecoregions have the greatest proportion of at-risk stocks. As with chum, a grouping of stocks classified at moderate extinction risk is centered around the Sea of Okhotsk (17–21) and western Bering Sea (23–24).

Masu

As with the other species, extinction risks are greatest at the southern edge of the species' range. Most or all masu stocks are classified at high risk of extinction in the southern Russian Far East (in the Russian portion of ecoregion 5 and 6–7). No masu extinctions are reported within Russia. In ten ecoregions, we classified all masu stocks at low risk of extinction.

Pink

Except in the extreme southern margins of their range, pink salmon generally exhibit the lowest level of risk of the seven salmon species. Within seventeen ecoregions, we classified all pink salmon stocks at low extinction risk. There are no pink salmon extinctions reported for the Asian Pacific. In North America, at least one stock is extinct in each of the five ecoregions. Pink salmon have been extirpated in the Klamath River (50) and Sacramento-San Joaquin River (51) ecoregions in North America.

Sockeye

Sockeye are at greatest risk toward the southern edge of their range in North America. At least some proportion of the sockeye stocks within all North American ecoregions are classified at a moderate or high risk of extinction, and eight North American ecoregions have at least one extinct sockeye stock. All stocks have been extirpated in the Klamath River (50) and Sacramento-San Joaquin River (51) ecoregions. In five Asian ecoregions, 100% of the sockeye stocks are categorized at low risk of extinction.

Steelhead

Steelhead have not been entirely eliminated from any of the ecoregions assessed. At least one stock has been extirpated in each of the seven North American ecoregions. We classify all steelhead stocks at low risk of extinction in three ecoregions (23, 40–41) and in the U.S. portions of the southeast Alaska ecoregion complex (36, 38–39).

Threats to Salmon Stocks

In the ecoregions for which we have stock-status information, we assessed the relative risks posed by various threats:

- Habitat destruction: Most pervasive threat in North America

- Over harvest: Most widespread threat in the Russian Far East, most common threat across the Pacific Rim

- Poaching: Major threat in the Russian Far East

- Hatcheries: Longtime threat in North America, emergent threat in the Russian Far East

- Logging and mining: Longtime threat in North America, emergent threat in the Russian Far East

- Oil and gas development: Primary threat in the southeast Sakhalin Current

- Dams: Substantial impact in the Sacramento-San Joaquin and Columbia River Basins, limited impact in the Russian Far East (cost and low-relief topography limits hydroelectric development)

Discussion

Our results describe a general latitudinal pattern of increasing extinction risk along the southern margins of salmon distribution in both Asia and North America. A number of factors, discussed below, contribute to this pattern. These include human population trends, edge-of-range effects, Pacific Decadal Oscillation, market pressures and economic structure, and the effects of hatcheries. Underscoring the impact of these combined factors is the fact that the gradient is consistent in spite of vast differences in geology and climate in North America and Asia (which would otherwise point to varying extinction risks in the East and West).

Human Population

In both Asia and North America, human population density increases with decreasing latitude (Figure 3). The trend is of course not endemic to the Pacific Rim; people tend to settle in milder climates, where it's easier to live, where resources are richer, and where climatic hardships are less burdensome. The results including habitat loss, pollution, and the depletion of water resources have deleterious effects on salmon stocks. Furthermore, landscape alteration for homes, agriculture, and natural resource extraction (e.g., timber harvest, oil and gas development, mining) have strongly negative cumulative effects on essential salmon spawning and rearing habitat (Spence, et al., 1996; Committee on Protection and Management of Pacific Northwest Anadromous Salmonids, 1996). As a result, there is a high correlation between human population density and the pattern of greater risk of salmon extinction in the more southerly latitudes of the Pacific Rim. Human population as a risk factor was captured in our risk survey as urbanization. Note that the spatial extent of urbanization risk for Kamchatka is distorted through characterization at an ecoregion scale (23; Bering Slope/ Kamchatka Current).

Edge-of-Range Effects

The levels of extinction risks for salmon across the North Pacific parallel the natural gradients in abundance often observed at the edge of a species' distribution (Brown, 1995). Commercial catch data for coho salmon, for example, indicate maximum concentration of the species in British Columbia, with a gradient outward from the center of its distribution to lesser concentrations (Fredin, 1980). This pattern likely coincides with environmental conditions that are ideal for coho in British Columbia but which become increasingly unfavorable for production and survival as the fish travel away from the distribution epicenter. Species are more vulnerable at the edges of their range and are more susceptible to the impacts of human population. Specifically with regard to coho, it follows that stocks that once thrived in California and Oregon are now at high risk or extirpated, as this heavily impacted coastline is on the perimeter of the species' range.

Of course, salmon do not have identical geographic ranges across the Pacific Rim, nor do they exhibit simple latitudinal clines in abundance. For example:

- Pink salmon stocks are widely distributed across sixty ecoregions on both sides of the Pacific Rim with peak abundance centered at high latitudes, although the center of their range is in Asia.

- Masu salmon occur in relatively small populations confined to eighteen Asian ecoregions.

- Chinook salmon abundance is shifted toward North America. Kamchatka is the only jurisdiction in the Asian Pacific with large populations.

In spite of these differences, however, pink, masu, and chinook all exhibit diminished populations at the southern reaches of their range, however wide or small their ranges might be. Nonetheless, we must note that because abundance epicenters are not identical, we cannot expect extinction risks to be simple mirror images on either side of the Pacific Rim.

Chinook are notable because they do not exhibit edge-of-range effects. Chinook salmon abundance is positively correlated throughout its range to large river size, independent of latitude. Major rivers near the limits of chinook distribution, such as the Sacramento-San Joaquin, historically supported population levels similar to stocks from large rivers in the middle of the range (Healey, 1991). This historic pattern of high productivity would explain why chinook stocks in the Sacramento-San Joaquin ecoregion are not at risk to the extent that latitude alone might suggest.

Pacific Decadal Oscillation

Changes in the North Pacific Ocean and atmosphere produce out-of-phase oscillations between the Gulf of Alaska and the ocean off Washington, Oregon, and California (Beamish and Bouillon, 1993; Francis and Hare, 1994; Francis, et al., 1996) and these have been shown to correlate with variations in salmon production. Although the theory of Pacific Decadal Oscillation (PDO) is relatively new, scientists believe the phenomenon is an old one: recent paleo-oceanographic studies have documented climatic variations with basin-wide effects on the North Pacific Ocean and its salmon, anchovy, and sardine populations long before fisheries or other anthropogenic factors had any significant influence (Finney, et al., 2000; 2002). A climatic regime shift in the late 1970s was a prime example of PDO; it was accompanied by record salmon harvests in the Gulf of Alaska region and in Asia and the rapid decline and subsequent listing of many Pacific Northwest stocks under the U.S. Endangered Species Act (Hare, et al., 1999). Even more recently, chinook and coho stocks, both naturally reproducing and of hatchery origin, increased dramatically in the early 2000s, coincident with a possible new, cooler climatic regime for the North Pacific Ocean. It is important to note that climate-driven cyclic variability ought not be immediately interpreted as species recovery or rebound; such a conclusion would be premature. PDO can mask trends occurring independent of climate variability and must be taken into account.

Harvest

Many cultural factors such as fishing practices, politics, industrial capabilities, and culinary customs inform and motivate harvest trends. In Russia, salmon eggs, or caviar, is highly prized, and pink and chum caviar is produced legally and illegally in great quantities. Historically Russian (including Soviet era) and Japanese fishing fleets targeted abundant Russian chum, pink and sockeye (Mandrik, 1994; Shepard, et al., 1985a). In the southern Russian Far East (notably ecoregions RK, 7 and 14; Zolotukhin, 2002), for example, edge-of-range effects play a role in the ranked risk of endangerment for chum and pink salmon. These stocks continued to be the prime commercial species targeted in the post WWII period.

In eastern Pacific ecoregions 47 through 53, coho and chinook are the high productivity stocks: populations have been badly hit as a result of human impacts (over harvest and habitat depletion), as well as natural impacts (natural productivity cycles). The majority of coho and chinook stocks in the contiguous United States are either listed under the U.S. Endangered Species Act or are candidates for listing. The once mighty Columbia River has seen its fishing days reduced to fall and winter seasons, the hatchery driven coho fishery is the most significant commercial fishery left in the Columbia River. Sockeye, once abundant in the Columbia River, has all but disappeared in the contiguous United States. Pink and chum populations, equally harvested throughout the Pacific, disappeared from the Columbia River southward on the eastern side of the Pacific decades ago because they were on the edge of their ranges.

Government Oversight

Not surprisingly, the United States and Canada have had a much heavier hand in monitoring and environmental protection than Russia. On the eastern side of the Pacific, for example, the now-endangered coho is protected under endangered species laws. The situation is entirely different on the western side of the Pacific Rim, where fishing has traditionally been run by massive cooperatives. From the 1930s through the 1980s, all industrial fisheries in Russia were controlled by the government. In the 1950s, personal use fisheries were banned; when licensed fisheries were opened in the 1990s, the market became a free for all (although officially access was limited). Today in the Amur, for example, fall chum harvest is as much as six times greater than is officially reported; elsewhere in Russia, actual harvest is at least twice as much as is reported (S. F. Zolotukhin, personal communication, Khabarovsk TINRO, Khabarovsk, Russia; Shuntov, 2002).

In addition, poaching for salmon roe is rampant and has taken a grave toll on all species, particularly chum, sockeye, coho, masu, and chinook. The result: rankings of moderate or greater risk of extinction for most species in Northeastern Russia.

One final note on government oversight: It is noteworthy that although the same poaching pressures are present on Sakhalin Island, local experts ranked all but two chum salmon populations at low risk of extinction. This fact demonstrates a clear difference in risk perception across the jurisdictions of the Russian Far East, a clear weakness in our best-expert-judgment approach.

Industry Characteristics

On the western side of the Pacific, fisheries traditionally have been communal ventures, characterized as cooperatives and/or government-run; using gear, such as beach seines and net traps that require large crews, which are most efficient for targeting smaller schooling fish that return in aggregates like pink and chum. The ethos of the North American fisherman, however, has been quite different: entrepreneurial, individualized, and market-driven. The gear was created to serve this function: boats were merchant-scaled, trollers geared toward hunting solo coho and chinook, or purse seiners and gillnetters geared toward capturing schooling pink, chum, and sockeye.

Political pressures in the 1920s through the 1950s favored small operators; therefore, highly labor-efficient gear, such as fish wheels or Alaska-style net traps operated by cannery operators or other private landowners, were eliminated from North American fisheries.

Hatcheries

In response to cyclical patterns of over harvest and declines in natural production, politicians most often respond with artificial production— hatcheries (Augerot, 2000; Lichatowich, et al., 1999). Japan and the U.S. Pacific Northwest introduced the concept in the mid nineteenth century; by the 1960s, the hatchery programs were entrenched (Kaeriyama, 1999). British Columbia and Alaska dabbled in hatcheries in the first half of the twentieth century but rejected artificial production as inefficient, opting for natural production in unaltered salmon rivers. These regions did not resume large-scale hatchery production until the 1970s, when natural variability in productivity, over harvest, and habitat alteration combined to produce declining harvests and political pressure for hatcheries. Despite a legacy of hatcheries on Sakhalin Island dating to the Japanese presence between 1907 and 1945, most of the Russian Far Eastern regions did not adopt artificial production strategies until the 1980s (Augerot, 2000). In most of the Russian Far East, freshwater habitat remains productive and the major limiting factors are harvest (M. Korolev, personal communication, Sevvostrybvod, Petropavlovsk-Kamchatsky, Russia) and natural variability (Radchenko and Rassadnikov, 1997).

Conclusion

In spite of contrasting environments, resource management histories, and cultural practices among jurisdictions of the Pacific Rim, a common ecological pattern emerges when salmon stock status is viewed at the entire North Pacific scale. We conclude that the latitudinal pattern of salmon decline is a major feature of the North Pacific ecosystem, defined by its salmonid biogeography, oceanic and climatic processes, and cultural interactions with salmon. Basin-wide climate variability and emergent patterns of climate change present serious implications for both oceanic and freshwater productivity. Low-productivity salmon stocks at the southern edges of the range (e.g., coho, masu, and steelhead) may be a bellwether for global warming induced climate effects throughout the North Pacific Rim (Welch, 1998).

As Nehlsen, et al. (1991) discovered in their study of status of salmon in the U.S. Pacific Northwest, stepping back from local details and examining salmon over a panoramic geographic area can be important in recognizing widespread inter-related problems with salmon and their ecosystems that may seem isolated and unrelated if viewed from a narrower spatial perspective.

Recommendations

This article represents a first step toward understanding the status of salmon and steelhead around the North Pacific Rim and the threats contributing to their decline. The trend observed (increased risk of extinction with decreased latitude) represents a serious threat to future salmon populations. Clearly more study is warranted, and we outline that path as follows.

First, salmon managers and scientists need to establish a standardized framework for monitoring salmon populations around the North Pacific Rim; in doing so, we must agree on a common census unit (e.g., stock, evolutionarily significant unit) and refer consistently to this unit throughout this vast region. Such an agenda is predicated upon sufficient budgetary momentum and confidence, and therefore regional and federal legislatures must be fully supportive. Second, we recommend the development of a trans-Pacific monitoring design, using salmon ecoregions as the design infrastructure. Third, the international salmon community will need to devise and adopt common abundance monitoring protocols that have been tested for appropriateness with regard to particular analytic tasks and field conditions. Fourth, we must identify a representative set of river basins across ecoregions in which to conduct routine monitoring and quantitative analysis. Furthermore, as these efforts proceed over the next few years and as our understanding of Pacific salmon marine migrations improves, we must revisit and modify our existing ecoregions to more accurately reflect stock boundaries and to ensure that they are biologically meaningful.

At the very least, the stock-status reviews reported in this article should be conducted at regular intervals to provide information on the trend of salmon stock status around the North Pacific Rim and to evaluate salmon stocks in Alaska, Japan, and Korea. A regular review of salmon stock status around the North Pacific Rim for all salmon ecosystems would be a useful tool for informing both resource managers and the public about the global status of anadromous salmon stocks and the implications of human activities.

On a parallel track, we must improve our understanding of the interactions between climate variability and salmon productivity in salmon ecoregions. Optimally we should improve our understanding of migration pathways and ocean rearing areas for specific stocks as a functional validation of our salmon ecoregions. Improved understanding of the interaction among climate variability, harvest, hatcheries, and habitat management will allow us to manage fisheries, artificial productivity, and land use in support of long-term salmon biodiversity and sustainability.

Acknowledgments

The authors would like to thank Stan Gregory and Mike Unsworth at Oregon State University for their vision and assistance in this work.

The following colleagues attended the North Pacific Salmon Workshop at Oregon State University, Corvallis, Oregon, May 4–6, 1999, and their input into this research was invaluable. From Canada: Kim Hyatt, Science Branch, Pacific Biological Station, Canada Department of Fisheries and Oceans. From Japan: Takemi Ichimura; Yutaka Okamoto, Okamoto International Affairs Research Institute. From Russia: Igor Chereshnev, Institute of Biological Problems of the North, Far East Branch, Russia Academy of Sciences; Elena Eronova, Khabarovsk Branch TINRO; Mikhail Korolyov, Sevvostrybvod; Evgeny Muzurov, Sevvostrybvod; Ksenia Savvaitova, Biology Faculty, Moscow State University; Anatoly Semenchenko, TINRO-Centre; Alexander Zhulkov, SakhNIRO; Sergey Zolotukhin, Khabarovsk TINRO. From the United States: Mike Beltz, The Ecology Center; Greg Bryant, NOAA Fisheries; Stanley Gregory, Oregon State University; David Gordon, Pacific Environment; Tim Haverland, Commercial Fish Division, Alaska Department of Fish and Game; David Heller, U.S. Forest Service Region 6; David Hulse, University of Oregon; Paul McElhany, NOAA Fisheries; Willa Nehlsen, U.S. Fish and Wildlife Service; Tom Nickelson, Oregon Department of Fish and Wildlife; Ron Nielsen, Pacific Northwest Research Station, U.S. Forest Service; William Pearcy, Oregon State University; Guido Rahr, Wild Salmon Center; Dorie Roth, Ecotrust; Alan Springer, University of Alaska—Fairbanks; Ed Weiss, Habitat Division, Alaska Department of Fish and Game. Additional data contributors not attending the workshop included: Dave Albert, Ecotrust-Canada; Leon Khorevin, SakhNIRO; Evgeny Korotkov, Sakhalinrybvod; Kirill Kuzishchin, Moscow State University; Sergey Putivkin, Magadan TINRO; Rich Lincoln,

Department of Fish and Wildlife, Washington; Vladimir Radchenko, SakhNIRO; Alexander Rogatnykh, Magadan TINRO; Mikhail Skopets, Wild Salmon Center and Institute of Biological Problems of the North, Russian Academy of Sciences, Valentina Urnysheva, Kamchatrybvod.

References

Augerot X. 2000. An environmental history of the salmon management philosophies of the North Pacific: Japan, Russia, Canada, Alaska, and the Pacific Northwest United States. Department of Geosciences, Oregon State University, Corvallis.

Augerot Xanthippe. 2005. Atlas of Pacific salmon. University of California Press.

Baker T.T., Wertheimer A.C., Burkett R.D., Dunlap R., Eggers D.M., Fritts E.I., Gharrett A.J., Holmes R.A. & Wilmot R.L. 1996. Status of Pacific salmon and steelhead escapements in southeastern Alaska. Fisheries 21(10):6–18.

Barber R.T. 1988. Ocean basin ecosystems. Pages 171–193 in Pomeroy L.R., Alberts J.J., editors. Concepts of ecosystem ecology: a comparative view. Springer-Verlag, NY.

Beamish R.J. and Bouillon D.R. 1993. Pacific salmon production trends in relation to climate. Canadian Journal of Fisheries and Aquatic Sciences 50:1002–1016.

Brown J.H. 1995. Macroecology. University of Chicago Press, Chicago, IL.

Cederholm C. Jeff, Johnson D.H., Bilby R.E., Dominguez L.G., Garrett A.M., Graeber W.H., Greda E.L., Kunze M.D., Marcot B.G., Palmsano J.F., Plotnikoff R.W., Pearcy W.G., Simenstad C.A. & Trotter P.C. 2000. Pacific salmon and wildlife—ecological contexts, relationships, and implications for management.

Committee on Protection and Management of Pacific Northwest Anadromous Salmonids, Board on Environmental Studies and Toxicology, Commission on Life Sciences. 1996. Upstream: salmon and society in the Pacific Northwest. National Academy Press, Washington, D.C.

Finney B.P., Gregory-Eaves I., Douglas M.S.V. & Smol J.P. 2002. Fisheries productivity in the northeastern Pacific Ocean over the past 2,200 years. Nature 416:729–733.

Finney B.P., Gregory-Eaves I., Sweetman J., Douglas M.S.V. & Smol J.P. 2000. Impacts of climatic change and fishing on Pacific salmon abundance over the past 300 years. Science 290:795–799.

Francis R.C., Hare S.R. 1994. Decadal-scale regime shifts in the large marine ecosystems of the Northeast Pacific: a case for historical science. Fisheries Oceanography 3:279–291.

Francis R.C., Hare S.R., Hollowed A.B. & Wooster W.S. 1996. Effects of interdecadal climate variability on the oceanic ecosystems of the NE Pacific. Fisheries Oceanography 7:1-21.

Fredin R.A. 1980. Trends in north Pacific salmon fisheries. Pages 59–120 in McNeil W.J., Himsworth D.C., editors. Salmonid Ecosystems of the North Pacific. Oregon State University Press, Corvallis.

Gende Scott M., Edwards Richard T., Willson Mary F. & Wipfli Mark S. 2002. Pacific salmon in aquatic and terrestrial ecosystems. BioScience 52(10):917–928.

Gresh Robert T., Lichatowich Jim & Schoonmaker Peter. 2000. An estimation of historic and current levels of salmon production in the northeast Pacific ecosystem. Fisheries 25:15–21.

Hare S., Mantua N. & Francis R. 1999. Inverse production regimes: Alaska and West Coast Pacific salmon. Fisheries 24(1):6–14.

Healey M.C. 1991. Life history of chinook salmon (Oncorhynchus tshawytscha). Pages 311–393 in Groot C., Margolis L., editors. Pacific salmon life histories. UBC Press, Vancouver, Canada.

Huntington C., Nehlsen W. & Bowers J. 1996. A survey of healthy native stocks of anadromous salmonids in the Pacific Northwest and California. Fisheries 21(3):6–13.

Ianovskaia N.V., Sergeeva N.N., Bogdan E.A., Kuydriavtseva A.V., Kalashnikova I.K., Romanova E.A., Gritsenko O.F., Vronskii B.B., Efanov V.N., Kostarev V.L., Kovtun A.A., Roslyi I.U.S. & Pushkareva N.F. 1989. Ulovy tikhookeanskikh lososei 1900–1986 gg. VNIRO, Moscow, Russia.

INPFC [International North Pacific Fisheries Commission]. 1979. Historical catch statistics for salmon of the North Pacific Ocean. International North Pacific Fisheries Commission Bulletin 39:1–166.

IUCN 2001. IUCN Red List categories and criteria: Version 3.1. IUCN Species Survival Commission, IUCN, Gland, Switzerland and Cambridge, UK. ii + 30 pp.

Kaeriyama M. 1999. Hatchery programs and stock management of salmonid populations in Japan. Pages 153–167 in B. R. Howell, E. Moksness, and T. Svasand, editors. Stock enhancement and sea ranching. Blackwell Science.

Karpenko V.I. 1998. Rannii morskoi period zhizni tikhookeanskikh lososei [The early sea life of Pacific salmons]. VNIRO Publishing, Moscow.

Lichatowich J., Mobrand L. & Estelle L. 1999. Depletion and extinction of Pacific salmon (*Oncorhynchus spp.*): a different perspective. ICES Journal of Marine Science 56:467–472.

Mandrik A.T. 1994. Istoriia rybnoi promyshlennosti rossiiskogo Dal'nego Vostoka. (50-e gody XVII v.–20-e gody XX v.). Dal'nauka, Vladivostok, Russia.

Nehlsen W., Williams J.E., Lichatowich J.A. 1991. Pacific salmon at the crossroads: stocks at risk from California, Oregon, Idaho, and Washington. Fisheries 16(2):4–21.

O'Neill R.V., DeAngelis D.L., Waide J.B. & Allen T.F.H. 1986. A hierarchical concept of ecosystems. Princeton University Press, Princeton, NJ.

Pearcy W.G. 1992. Ocean ecology of north Pacific salmonids. Washington Sea Grant Program, University of Washington Press, Seattle.

Radchenko V. 1998. Historical trends of fisheries and stock condition of Pacific salmon in Russia. North Pacific Anadromous Fish Commission Bulletin 1:28–37.

Radchenko V.L., Rassadnikov O.A. 1997. Tendentsii mnogoletnei dinamiki zapasov aziatskikh lososei i opredeliaiushchie ee faktory. Izvestiia TINRO 122:72–94.

Ricker W.E. 1972. Hereditary and environmental factors affecting certain salmonid populations. Pages 19–160 in Simon R.C., Larkin P.A., editors. The stock concept in Pacific salmon. University of British Columbia, Vancouver.

Seong K.B. 1998. Artificial propagation of chum salmon (*Oncorhynchus keta*) in Korea. *North Pacific Anadromous Fish Commission Bulletin* 1:375–379.

Shepard M.P., Shepard C.D. & Argue A.W. 1985a. Historic statistics of salmon production around the Pacific Rim. Canadian Manuscript Report of Fisheries and Aquatic Sciences 1819:1–297.

1985b. Long-term trends in the contributions of salmon from different geographic areas to the commercial fisheries of the North Pacific. Canadian Technical Report of Fisheries and Aquatic Sciences 1376:1–52.

Sherman K. and Duda A.M. 1999. Large marine ecosystems: an emerging paradigm for fishery sustainability. Fisheries 24(12):15–26.

Shuntov V.P., Dulepova E.P. & Volvenko I.V. 2002. Sovremennyi status i mnogoletniaia dinamika biologicheskikh resursov dal'nevostochnoi ekonomicheskoi zony Rossii. Izvestiia TINRO 130:3–11.

Slaney T.L., Hyatt K.D., Northcote T.G. & Fielden R.J. 1996. Status of anadromous salmon and trout in British Columbia and Yukon. Fisheries 23(10):20–35.

Spence B., Lomnicky G., Hughes R.M. & Novitzki R. 1996. An ecosystem approach to salmonid conservation. Report TR-4501-96-6057. ManTech Environmental Research Services Corporation, Corvallis, OR.

Stanford J.A., Lorang M.S. & Hauer F.R. 2005. The shifting habitat mosaic of river ecosystems. Verh. Internat. Verein. Limnol. 29(1):123–136.

Strahler A.N. 1957. Quantitative analysis of watershed geomorphology. Transactions of the American Geophysical Union 38:913–920.

Suslov S.P. 1961. Physical geography of Asiatic Russia. Translated by N.D. Gershevsky, edited J.E. Williams. W.H. Freeman and Company, San Francisco, CA.

Ware D.M., McFarlane G.A. 1989. Fisheries production domains in the northeast Pacific Ocean. Pages 359–379 in Beamish R.J., McFarlane G.A., editors. Effects of ocean variability on recruitment and an evaluation of parameters used in stock assessment models. Canadian Special Publication in Fisheries and Aquatic Sciences 108.

Welch D.W., Ishida Y., Nagasawa K. & Eveson J.P. 1998. Thermal limits on the ocean distribution of steelhead trout Oncorhynchus mykiss. Bulletin of the North Pacific Anadromous Fish Commission 1:396–404.

Zolotukhin S.F. 2002. Current threats to the salmon biodiversity in rivers of the Khabarovsk Territory and the Amur River. Pages 63–64 in Abstracts; First International Symposium on Biodiversity of Freshwater Fishes of the Amur River and Adjacent Territories. 29 October–1 November, Khabarovsk Branch Pacific Research Fisheries Center, Khabarovsk, Russia.

Historic Voyage
as Catalyst for Inspiring Change

Ann Melinda Bell
Outdoor Recreation Planner, U.S. Fish and Wildlife Service

For *Hōkūleʻa*, a replica of an ancient voyaging canoe, navigator Nainoa Thompson coined the phrase "Navigating Change" in order to implant inspiration in the hearts and minds of Hawaiʻi's youth to take better care of their island home. Ultimately, it was about instilling hope and a culturally based value of responsibility in our younger generation. In 2001, a partnership of organizations began to develop tools based on the Navigating Change vision to complement *Hōkūleʻa's* voyage to the Northwestern Hawaiian Islands.

During the voyage, more than 800 students were involved via satellite teleconferencing conversations with crewmembers. The voyage generated almost 2,500 column inches of newspaper coverage and nearly two hours of television news coverage. For the past three years, more than 200 teachers have been directly involved in implementing a comprehensive "Teacher's Guide to Navigating Change."

The end of a voyage turned into a new beginning for Navigating Change. A student-driven community day was held in May 2005, to honor the cumulative conservation learning and work of hundreds of student with more 5,000 people in attendance as *Hōkūleʻa* sailed into Kailua Bay. In addition, the Harold K. L. Castle Foundation funded a half-time NCEP position to help steer the vision of Navigating Change into the future.

When Nainoa Thompson was a boy growing up in Hawaiʻi , the reefs were teeming with pāpio (juvenile Jack fish), goatfish, and āholehole (Hawaiian flagtail). Mullet drifted in so thickly they looked like the reflection of dark clouds on the water. Nainoa's keen sense of the ocean world around him sharpened as he dove for lobsters, surfed the waves, and learned to fish beside men who filled their boats full to feed all the community. Back then, Maunalua Bay provided islanders with a critical lifeline, connecting them directly to the teeming source of their livelihood. Both nursery and spawning ground, the bay gave shelter and food sources to millions of native fish that in turn nourished the bodies and souls of the islanders who lived in the lee of the bay's *ahupua'a* (a traditional land and community division running from mountain to sea). It also served as an outdoor classroom, and the lessons Nainoa learned there guide him to this day as he navigates *Hōkūle'a*, the replica Polynesian voyaging canoe, through the ocean waters of the twenty-first century.

"Whether people want to recognize it or not, we are connected to our natural environment," Nainoa says. Many people living in Hawai'i today don't understand how their disconnection from their surroundings affects their well-being. "What we do to the land and sea, we do to ourselves. So if we take care of even the smallest portion of land or ocean or the smallest creature, we take care of ourselves."

Figure 1 *Hōkūle'a* sails past Nihoa Island in the Hawaiian Islands National Wildlife Refuge. Photo by Na'alehu Anthony

Things have changed drastically in Maunalua Bay. No longer the rich haven Nainoa experienced fifty years ago, a private marina, shopping centers, and condos line the adjacent shoreline of what was once an enormous ancient fishpond feeding into the bay. While natural tidal flows brought in fish that provided sustenance for the entire community, today this passageway is a dredged, silt-covered thoroughfare that provides access for boats of all kinds, including every imaginable kind of boat or water toy. The adrenaline rush from fast boats and video games has replaced that gained from exploring and experiencing the natural world. No longer can you teach a child how to find lobsters. Lobster populations have dwindled to the point that diving is not worth the effort. Although patches of coral in deeper areas are still alive with sponges, algae, and invertebrates, the overall biomass of fish has dropped by 80% during the last fifty years. The large schools of reef fish are gone and the sizes of individual fish are greatly diminished.

While natural resource agencies, organizations, and community-based initiatives have struggled for years to protect the remnants of our native ecosystems, they have had little success in gaining much political support for the importance of their efforts. The total state funding for natural resource protection remains tragically low, less than 1% of the state budget. Degraded resources are accepted as normal, alien species are often accepted as Hawaiian, more and more native species are threatened with extinction, and the potential for negative impacts on human health are increasing. "A child today sees a world that is substandard and degraded," says Nainoa. "Through the eyes of a child, this picture is the picture of what is healthy."

Nainoa's passion is to reconnect people to their world. He believes that learning to care about place requires teaching children healthy traditional values and demonstrating that actions have real solutions. "When a child loses the capacity to understand or care about place, a disconnect occurs," Nainoa says. "If the gap is present now, it is only going to get bigger in the future. We must help students reconnect by providing opportunities that reawaken their observational skills and help them understand the value of nurturing their own spirituality and physical well-being through taking care of their place."

Navigator Nainoa Thompson envisioned reconnecting people with place by sailing the *Hōkūleʻa* among the wild and protected Northwestern Hawaiian Islands. Calling his idea "Navigating Change," he wanted to, "Bring the beauty of Earth's rare wildlife to living rooms and classrooms to create an awareness of the difference between where nature is protected and what happens when it is not."

Primarily an educational program echoing his father's vision, Navigating Change was designed to inspire people to *mālama* (take care of) their native land. Thompson wants people to understand that to live well and be healthy, your ocean must also be healthy, and that for your ocean to be healthy, it must mirror a healthy land. Navigating Change provides an opportunity to show people what they have lost and what they need to do to reverse the damage. At the core of Navigating Change is *Hōkūleʻa*, a modern-day reincarnation of a double-hulled sailing vessel that has accomplished almost inconceivable navigational feats, using science built upon a foundation of ancestral knowledge.

A thousand years before Columbus approached North America, Polynesians were sailing across the Pacific. They voyaged to Hawaiʻi first from the Marquesas around 1,900 years ago, and around 1200 A.D. a second group of Polynesians headed north from Tahiti, which lies approximately 1,000 miles to the southwest of the Marquesas. These long-distance voyages were perhaps made to seek more abundant island resources, to escape oppression due to societal conflicts, or perhaps for exploration purposes. Over long days and nights on the open sea, Polynesians continued to hone their traditional practice of wayfinding by implementing their vast knowledge of the stars, winds, birds, and waves that guided them to Hawaiʻi.

In the late 1970s, several hundred years after such long-distance voyaging activity ceased, a group including anthropologist Ben Finney, artist Herb Kawainui Kane, and waterman Tommy Holmes designed and facilitated the construction of a modern-day voyaging canoe modeled after ancient double-hulled sailing vessels that had platforms lashed to the crossbeams. The canoe was named *Hōkūleʻa* after the star of glad-ness, which is the Hawaiian name for Arcturas, the zenith star that marks the islands of Hawaiʻi. To find a navigator skilled in the ancient ways was no easy feat, but eventually Mau Piailug, from the tiny island of Satawal in Micronesia, agreed to share his ancestral knowledge. His teachings inspired young Hawaiians like Nainoa Thompson, Bruce Blankenfeld, Shorty Bertlemann, and Chad Babayan to spend years learning the ways of the waves, wind, and stars.

Over the last thirty years, *Hōkūleʻa* has sailed more than 100,000 miles across the Pacific. On its last long-distance voyage in May 2004, *Hōkūleʻa* followed in the wake of Hawaiʻi's ancestors to the Northwestern Hawaiian Islands. These islands, all but one of which are within the Hawaiian Islands and Midway Atoll National Wildlife refuges, extend along the northern half of the Hawaiian archipelago, reaching more than 1,200 miles northwest of the main Hawaiian Islands. This string of atolls, reefs, and islets embodies the definition of wilderness, and in 1974 most of the emergent land in this area was proposed as wilderness under the National Wilderness Preservation Act. Further protection will be

Figure 2 Navigator, Nainoa Thompson, on satellite phone while U.S. Fish and Wildlife Service employee, Ann Bell, helps coordinate communication with students back home. Photo by Dr. Randall Kosaki

afforded to the surrounding state waters when the governor of Hawai'i signs regulations creating the State Marine Refuge.

Extending out fifty miles, the marine ecosystem is being studied for a potential designation as the country's largest national marine sanctuary. This coral reef ecosystem is believed to be one of the last of its kind, alive with vestiges of marine and island wildlife that have long since disappeared from the main Hawaiian Islands. It is also one of the last places of its size on the planet in which dominant large marine predators live in concert with a diverse entourage of coral, fish, and birds, indicating a healthy, balanced system. The numbers and varieties of species are exceptional; thousands of species exist in the Northwestern Hawaiian Islands and nowhere else on the planet.

But because of their remote location and fragility, the Northwestern Hawaiian Islands will rarely be able to accommodate visitors seeking to enjoy their solitude and primitive recreational opportunities. Instead the Fish and Wildlife Service is working with staff from NOAA and the State of Hawai'i to "Bring the place to the people, rather than the people to the place." Our challenge is to infuse in our audiences the spirit of wilderness found within these islands.

Hōkūle'a's eighteen-day voyage through the Northwestern

Figure 3 La Piétra, Hawai'i School for Girls Environmental Science Class discovers the difference between native and non-native algae in Maunalua Bay, Hawai'i. Photo by Jessica Carew

Hawaiian Islands was the culmination of almost three years of preparation. In the winter of 2002, Nainoa Thompson pulled together a partnership of agencies to create educational projects and products that coordinate with and support the Navigating Change voyage and vision. He directed the partnership group on a course that would impact the lives of hundreds of students and their families. The partnership included key educators from the Polynesian Voyaging Society, Bishop Museum, Hawaii Department of Land and Natural Resources, Hawaiʻi Department of Education, Hawaiʻi Maritime Center, National Oceanic and Atmospheric Administration (NOAA), National Fish and Wildlife Foundation, U.S. Fish and Wildlife Service, and the University of Hawaiʻi. The group was soon labeled as the Navigating Change Educational Partnership (NCEP).

In September 2002, Nainoa Thompson and several NCEP educators sailed on the NOAA vessel *Rapture* as part of the Northwestern Hawaiian Islands Reef Ecological Assessment and Monitoring Program. Concurrently, the Bishop Museum's Hawaiʻi Maritime Center opened a permanent Navigating Change interactive exhibit. The exhibit, funded by NASA and highlighting the research expedition, allowed NCEP educators to transmit almost real-time video segments via cutting-edge satellite technology to students visiting the museum. A regularly updated interagency website created on board the vessel, along with significant media coverage, generated increasing interest in these far flung islands. During this time, *Hōkūleʻa* was in dry dock undergoing extensive restoration as volunteers repaired dry rot, sanded, varnished, and carefully pulled lines taut.

By the spring of 2003, NCEP had developed a package of teacher resources, a Northwestern Hawaiian Islands map poster, and a series of five video modules focused on specific Navigating Change educational themes. In April 2003, *Hōkūleʻa* departed on a seven week statewide sail in which the canoe would visit most of the main Hawaiian Islands, allowing students to meet the canoe and learn about voyaging and Navigating Change first hand. A series of nine teachers' workshops were held in conjunction with this trip.

Hōkūleʻa was scheduled to set sail to the Northwestern Hawaiian Islands in late summer of 2003. Unfortunately, two weeks before departure, a threatening hurricane and a broken mast on the escort vessel delayed departure. With winter weather approaching, *Hōkūleʻa's* voyage was rescheduled for May 2004. Although disappointed, NCEP took advantage of the time to fine tune its educational products. In the fall of 2003, the State of Hawaiʻi's Department of Education, with assistance from NCEP, aired on Public Television a three-part series on Navigating Change via an interactive distance-learning science program. The teacher's guide was updated to incorporate the Department of Education's content and performance standards in science, social studies, language arts, and Na Honua Mauli Ola (Hawaiʻi guidelines for culturally healthy and responsive learning environments), so that teachers could easily incorporate the guide into their standardized curricula. This updated guide was reworked into a framework of digestible topical units that coordinated with the existing poster-sized map, video segments, photographs, Power Point presentations, the Hawaiʻi Maritime Center exhibit, and websites.

On May 23, 2004, *Hōkūleʻa* set sail for the Northwestern Hawaiian Islands. To improve interaction with the public, the twelve-member crew included a journalist, Jan TenBruggencate from the *Honolulu Advertiser,* and an education and ecological protocol officer, Ann Bell of the U.S. Fish and Wildlife Service. During the voyage, more than eighty classrooms and approximately 1,800 students were connected to the canoe's crew via satellite telephone. During the first two weeks of the voyage, daily forty-five-minute conversations allowed students from across Hawaiʻi, as far east as Maryland, and as far south as Samoa to ask questions and learn about the Northwestern Hawaiian Islands. TenBruggencate reported on day-to-day life on board the canoe, often in front-page articles, instilling a greater awareness in adults of the traditional Hawaiian value of *mālama,* caring for our land and sea. Three websites followed the voyage, posting extensive information along with journal articles by Dr. Cherie Shehata, and the public could track the canoe's daily position via a satellite tracking system.

The true measure of success is perhaps best told by the participants themselves. LaTitia A. McCoy, an eighth-grade teacher at Labadieville Middle School in Thibodaux, Louisiana, wrote:

"I cannot say enough about the resulting experiences that these students had the opportunity to be a part of! By integrating the Navigating Change project into all subject areas, the students were actively engaged in the connections that were being made. In science, the students learned about the ocean currents, trade winds, and navigation using the constellations above. In mathematics, tracking of the vessel was done using the longitude and latitude coordinates daily. In social studies, the geography of the islands was taught, and students learned in-depth information about the Northwestern Hawaiian Islands that most of them never knew existed before this project. Most students didn't even know that the fiftieth state of Hawaii consisted of more than one island when we began this project! The English/language arts teachers even became involved by exploring new vocabulary words that the students were exposed to. In the midst of the project, every eighth grade student could tell you what the Hōkūleʻa was, where and what the NWHI were, and how this event was to make an impact on their lives.

"I'm sure that the impact of this event will continue far longer than any of us can imagine, but some immediate signs that these young people absorbed the information that was being presented to them were evident in their responses to any question that was asked of them about the project. They responded with quick connections being made from Hawaiʻi being surrounded by water, and Louisiana being a coastal state. The erosion that takes place at an alarming rate is a concern for most south Louisiana residents, and these young people are aware of the problem and hope to slow the process in their lifetime. Protecting the ecosystem is a concern also, and hearing firsthand about endangered birds that were encountered through the voyage of the Hōkūleʻa brought the vision of a harmonious ecosystem to life for them.

"Cultural harmony is another issue that most young people here in Louisiana deal with on a daily basis. Hearing with their own ears (from Bruce) during one of the teleconferences that people of all races and ethnic backgrounds work together toward one common goal is an important asset for all crewmembers. It doesn't matter what the color of your skin is or where you were raised, only that we are all human beings and together we can make this world a better place to live in for the future. This was the overall feeling that the students at L. M. S. left with after completing their last teleconference. The feeling was overwhelming for me as a teacher to see these students absorbing this 'real-life' connection that was being made. This entire experience is one that no one at Labadieville Middle School will soon forget!

"You have impacted more than 120 students' lives in our school alone, not to mention all of the adults who read the local reports from our reporting media."

From Kilauea E. School on Kaua'i, Richard Larson said:

" … The experience of having the children speak with you on the canoe was the most significant event from the whole year, and it happened the day before school was out. It was a fitting celebration for the year.

"For many years I have used the voyaging canoe as a symbol for the year—the cooperation, the bringing together the knowledge of the past and the present, the unseen and the seen. With so much focus on standards and assessment, I have been able to integrate what I believe to be important into the daily activities and the curriculum. The values of *'ohana* (family), *aloha* (compassion), *kuleana* (responsibility) are just some of the cultural aspects that we use as part of the tapestry of our day, our year together."

The end of a voyage turned into a new beginning for Navigating

Change. A student-driven community day, organized by Learning Education Technology Academy, was held in May 2005, with more than 5,000 people in attendance as *Hōkūleʻa* sailed into Kailua Bay to honor the cumulative conservation learning and work of hundreds of students. Seven teachers who were previously involved in developing and field testing the Navigating Change teacher's guide in their classrooms were chosen to set sail in August 2005 on a NOAA ship to explore and produce lesson plans about the Northwestern Hawaiian Islands. In addition, the Harold K. L. Castle Foundation funded a half-time NCEP position to help steer the vision of Navigating Change into the hearts and minds of Hawaiʻi's children.

More than sixty students will spend the night on *Hōkūleʻa* during the fall of 2005, as it anchors in Maunalua Bay, a bay that nourished local families for hundreds of years and inspired Nainoa Thompson as a child. With *Hōkūleʻa* acting as a floating laboratory, students will create their own baseline studies of the coral reef, search the night skies and learn the art of wayfinding from Thompson, and experience the human lessons learned by working and sailing a voyaging canoe together. In these acts, the values of culture and science combine to show students that to help

Figure 4 Highly endangered Hawaiian monk seal (approximately 1,200 left) in the newly declared Papahānaumokuākea Marine National Monument (Northwestern Hawaiian Islands). Photo by James Watt

their crewmembers, their families, and the ocean, is to help all life become sustainable and healthy. As Thompson explains it, "No longer do we seek only the knowledge of how to voyage between islands. We seek lessons to carry home to our children—ways to inspire the present generation to love and preserve our Earth as a sanctuary for those who will inherit it."

Editor's Note

In September 2005, after a three and a half year public process, Hawaii's Governor Lingle established a State Marine Refuge in the NWHI that set aside all state waters surrounding the Hawaiian Islands National Wildlife Refuge and Kure Atoll as a limited access, no-take marine protected area. This created the largest marine conservation area in the history of the state, protecting 1,026 square miles of coral reefs from the shoreline to three miles offshore. To ensure similar protections at the national level, President Bush in June 2006 designated it as a Marine National Monument that now encompasses the islands and atolls of both the Hawaiian Islands and Midway Atoll National Wildlife Refuges and Kure Atoll out to fifty miles offshore. To honor the monument's cultural significance, in March 2007 the largest marine protected area in the world was renamed Papahanaumokuakea National Marine Monument. It encompasses nearly 140,000 square miles—more than 100 times larger than Yosemite National Park and larger than the collective land area of forty-six states. The island chain is home to more than 7,000 marine species, a quarter of which are found nowhere else on Earth.

Chapter 6

The Role of the Private Sector in Wilderness Conservation

"Clans"

~

Jean Anderson

Honorable Mention,
8th World Wilderness Congress Poetry Contest

Your mother would be a raven, your father
a whale. It comes to you while shoveling
foot upon foot of dry new snow, the third morning
of this in a land you were surely born to—
though in fact born Outside.

Your mother was raven-haired and mysterious,
secretive, frail, complex, a woman in Poe
and your father so large, so playful, alert but

gentle: born to surface and then go deep,
born to give himself.

Are we all woven of tangled and opposite
strand? You see it suddenly: the past
is the sky above us and the sea far beyond.
And you yourself, bending over your shovel,
you, again, are the earth.

The Sanbona Wildlife Reserve
Private Wilderness Area

Adrian Gardiner

CEO, The Mantis Collection

At the 7th World Wilderness Congress in Port Elizabeth, South Africa (2001), the Shamwari Game Reserve, situated in the Eastern Cape province of South Africa, declared the first private wilderness area in Africa. This was done by giving the Wilderness Foundation (South Africa) servitude over the 3,000 hectares (7,400 acres) proclaimed wilderness area. Furthermore a comprehensive management plan was drawn up for the wilderness management.

Shamwari Game Reserve is a private game reserve consisting of 20,000 hectares (49,000 acres); 15% of this area (3,000 hectares or 7,400 acres) has been declared a private wilderness area. This area falls within the Maputaland-Pondoland-Albany Global Biodiversity Hotspot, which enhances the conservation status of the wilderness area. There have, however, been both advantages and disadvantages in having a wilderness area on the reserve:

Disadvantages
- The 15% land area does not contribute significantly to the socio-economic benefits to the area

- Management with regard to predator, rhino and vegetation monitoring is hampered

- High-profile wildlife species entering the wilderness area become unavailable for tourist viewing

- Fence maintenance becomes very time consuming

Advantages
- The wilderness area enjoys the benefits of the high conservation priority
- The area acts as a control with respect to the tourism densities on the reserve as a whole
- Students, under privileged, and high-profile tourists are exposed to wilderness concepts, ideals, and ethics
- Has excellent marketing and PR potential
- Portrays the serious conservation message of Shamwari Game Reserve

As all aspects of a reserve make up the product, the wilderness area of Shamwari Game Reserve has over the past five years proven to be an essential and valuable addition in assisting in "Conserving a Vanishing Way of Life."

Having created the model, in conjunction with The Wilderness Foundation, for wilderness on private land through the Shamwari experiment the lack of other private landowners following the example can be seen as a failure. The Mantis Collection, however, is taking the private wilderness concept a step further by proclaiming a second wilderness area, under the custodianship of The Wilderness Foundation, on Sanbona Wildlife Reserve.

Sanbona Wildlife Reserve Wilderness Area

Sanbona Wildlife Reserve is a 54,000 hectare (133,400 acre) reserve in the Western Cape, South Africa. If falls within the Succulent Karoo Global Biodiversity Hotspot. It gives us great pleasure to declare 8,600 hectares (21,250 acres) or 16% of Sanbona a private wilderness area.

The Succulent Karoo boasts the richest succulent flora on Earth, as well as a remarkable endemism in plants, reptiles, and invertebrates. Sanbona is a semi-arid reserve, home to a variety of indigenous wildlife including predators and mega herbivores.

Conclusion

The Mantis Collection has shown its dedication to conservation and the wilderness concept by declaring the first two private wilderness areas in Africa. The following aspects emphasize the importance of these declarations:

- The process has been set for other private landowners to follow

- Combined, the two wilderness areas make up almost 12,000 hectares (29,650 acres)

- Between the two wilderness areas, all three of South Africa's Global Biodiversity Hotspots are represented (i.e., The Maputaland-Pondoland-Albany, Succulent Karoo, and Cape Floristic Region)

- The wilderness concept can now reach more people over a wider spectrum of life

The wilderness areas within Mantis have and always will have a symbiotically beneficial relationship with the collection.

Perhaps we, as the wilderness movement, have not marketed this achievement to its potential. If we consider wilderness to be the direction of the future, we should be directing and exploring possible private wilderness areas in a far more aggressive way.

Wildlands Philanthropy
An American Tradition

Tom Butler
President, Woodshed Communications

The year was 1926. Grace Wright, a student at Knoxville's Central High School, gave a nickel. Her classmate Herb Vesser chipped in fifty cents. They were among some 4,500 kids from east Tennessee who contributed to a fundraising drive then underway to purchase private lands for a new park: the Great Smoky Mountains National Park. The money raised by the children was a small fraction of the estimated $10 million price tag, but their enthusiasm gave a psychological boost to the park campaign for the Smokies, an extraordinarily scenic and biologically rich landscape.

Two years later, John D. Rockefeller, Jr. pledged $5 million to the effort, to match the collected donations of individuals, businesses, and the state legislatures of Tennessee and North Carolina. From school kids in Appalachia to the nation's preeminent philanthropist, Americans were swept up in the excitement of the early parks movement, which bubbled during the 1920s like Yellowstone's mudpots and geysers. The young National Park Service was a hotbed of activity. Park boosters across the country were organizing, garnering political support for their favored natural areas, a phenomenon that reflected both pride of place and a desire to seize market share.

Henry Ford's Model T had made automobile-based tourism a possibility for the average family, whetting the wanderlust of millions of Americans. At the time, aesthetics, recreation, and economic development were the primary arguments for protecting parks and other natural areas, although certainly many conservationists were dedicated to saving wild nature for its own sake.

The struggle to create Great Smoky Mountains National Park was nearly derailed later by The Great Depression, opposition from timber companies, and the challenge of buying thousands of individual properties to assemble into a national park, only the second in the eastern United States. But conservationists persevered, and prevailed; the park was fully authorized by Congress in 1934, dedicated by President Roosevelt in 1940, and today receives nearly 10 million visitors a year. It is a miracle, an intentional miracle, that in the midst of this long-settled landscape, a big, wild, beautiful place still exists, a refuge for nature and people. It would not exist if not for private initiative and funding: wildlands philanthropy.

For nearly 150 years, Americans have been setting aside some parts of the landscape from exploitation, initially focusing on the continent's natural spectacles. Today, our national and state parks, wilderness areas, wildlife refuges, and private nature preserves, such as those maintained by the Nature Conservancy and National Audubon Society, are among the nation's most beloved landscapes. The National Park System alone receives more than 250 million visits annually, but the creation stories of individual protected areas are little known. And despite the crucial role wildlands philanthropy has played in American conservation history, it is a largely uncelebrated and unstudied phenomenon.

During the twentieth century, some of America's most prominent families, with names like Rockefeller, McCormick, Mellon, and Dupont, used personal wealth to protect public values by purchasing private property and assuring its conservation as wildlife habitat. Today it seems that a wildlands philanthropy resurgence is underway. Examples on a grand scale include the unprecedented investments in biodiversity conservation made by high-tech titan Gordon Moore, and the purchase and transfer to public ownership of more than 600,000 acres of southern California desert lands, including inholdings in Joshua Tree National Park, made possible by a modest mathematician turned hedge fund man.

Continuing the parks creation legacy established by John D. Rockefeller, Jr. are Kristine and Doug Tompkins, former clothing company executives who have helped establish several new protected areas in South America, including two Yosemite-scale national parks along the Chilean coast. But private support for conservation is not solely the purview of successful entrepreneurs and Wall Street wizards. Anyone can help save natural areas, and Americans of every stripe are doing so. The explosion of the land trust movement, with more than 1,500 local and regional trusts now protecting land nationwide, represents the democratization of private conservation action and is a particularly hopeful trend.

These examples and the others to come suggest the crucial role that private funding and initiative have played in conservation history. One need not remember details of these stories to absorb three key points:

- Wildlands philanthropy is a great American tradition, which deserves to be better known and widely practiced

- That it is democratic and diverse, with room for everyone to participate

- That even as we promote it as an effective conservation tool, wildlands philanthropy by individuals should complement, never supplant, a strong public commitment to conservation, including adequate funding streams by governments

Wildlands philanthropy takes many forms. It may by practiced by volunteering for or donating to a local, grassroots group like the Highlands Nature Sanctuary in Ohio. With a cadre of committed volunteers, that organization has raised several million dollars to preserve some 2,000 acres, creating a ribbon of conservation lands along the Rocky Fork Gorge.

It may take the form of a gift to a regional organization, such as when Ann Down purchased a ranch on the east fork of the Salmon

River in Idaho and donated it to the Western Watersheds Project. Greenfire Preserve was born, and cattle were removed from 50,000 acres of federal grazing allotments attached to the property.

It may take the form of a gift to a national group, like when Mary Griggs Burke donated her family's wilderness retreat in northern Wisconsin to the Trust for Public Land (TPL), which then resold it to the Chequamegon National Forest. That transaction resulted in nearly 900 acres protected along Lake Namakagon and a pot of new funds for TPL to use protecting other imperiled land in the Northwoods.

Sometimes wildlands philanthropists eschew an intermediary and donate land directly to a state or federal agency for conservation, such as when Vermonter Joseph Battell gave the summit of his state's most beloved mountain, Camel's Hump, to the public for a state park in 1911.

Gordon McCormick, grandson of reaper inventor Cyrus McCormick, made a similar choice. When Gordon died in 1967, he willed the family's 17,000-acre tract of Michigan woods to the U.S.

Mary Wharton Preserve at Floracliff, near Lexington, Kentucky, U.S.A. Photo © Antonio Vizcaino/America Natural

Forest Service with the intent that it remain a wild forest. Congress subsequently designated it the McCormick Wilderness Area.

A relatively new avenue for wildlands philanthropy is the use of conservation easements, which may legally extinguish development rights on private land. The easement that sisters Annie, Abigail, and Emily Faulkner donated to the Maine Coast Heritage Trust will keep Norton Island forever wild.

Sometimes an unusual or threatened natural area will stimulate an outpouring of conservation concern and funding in a community where there was no precedent, as when the discovery of Kentucky's largest remnant of primeval forest led to the founding of the Kentucky Natural Lands Trust and a successful campaign to buy and preserve Blanton Forest.

Some wildlands philanthropists, particularly entrepreneurs accustomed to running their own businesses, choose to do their own conservation deals. Roxanne Quimby, the founder of the natural body care products company Burt's Bees, is a notable current example. Quimby has established a private foundation, which now holds title to some 50,000 acres of the Maine woods she has purchased for preservation, including property at the base of Mount Kineo that would have been developed for a dozen lakeside vacation houses had she not purchased the land.

Finally, some people choose to make their most significant conservation investments through estate planning. A $125 million gift to the Open Space Institute from the estate of Lila and DeWitt Wallace, the founders of *Reader's Digest*, has helped underwrite many conservation projects in the Hudson River Valley, including the creation of Sam's Point Preserve, part of an ambitious conservation corridor called the Shawangunk Ridge Greenway.

My purpose here, though, is not a survey of statistics or styles of conservation giving. I simply want to share stories. For a forthcoming book on wildlands philanthropy, photographer Antonio Vizcaino and I have been compiling stories of extraordinary places and the people who saved them. So let's take a brief tour across North America, stopping at just a handful of the thousands of natural areas protected through private conservation action.

Far out on the Alaska Peninsula, where a string of volcanoes rises from the Pacific, the Izembek National Refuge offers outstanding wildlife habitat. Here, caribou are kept wary by wolves, and seasonal concentrations of shorebirds defy enumeration. In the largest transfer of private land to public ownership in Alaska history, Izembek grew by 37,000 acres in 2002 when a native corporation sold property for addition to the refuge. The Richard King Mellon Foundation funded the project, which put it over 1 million acres conserved through its American Land Conservation Program.

On his travels through the Sea of Cortez, John Steinbeck once described how Espirtu Santo Island rises "high and sheer from the blue water." The uninhabited 23,000-acre island off the southern tip of Baja, Mexico, is now a federally protected natural area. A cohort of conservationists, including the American founder of an ecotourism company, Mexico's leading conservation philanthropist, and the David & Lucile Packard Foundation, saved it from subdivision and development.

Across the continent, where the Atlantic polishes square-edged granite into cobbles, George Bucknam Dorr exhausted his life's energy and family fortune creating the East's first national park. At the urging of Harvard President Charles Eliot, in 1901 Dorr helped found a nonprofit trust to conserve land on Maine's Mt. Desert Island. Eventually the group gave its holdings to the federal government for a national monument, which was upgraded to national park status in 1919. Some twenty-nine of the donated properties had been purchased by Dorr, who remained Acadia National Park superintendent until his death.

When Isaac Wolfe Bernheim was born in Germany in 1848, Jews were taxed but could not vote. The product of a society that institutionalized prejudice, even as a child he revered Thomas Jefferson's writing on liberty. After immigrating to the United States in 1867, he made his living as a peddler. Thanks to a growing nation's growing thirst for good Kentucky bourbon, within a few decades Bernheim Brothers distillery was a prominent Louisville business and Bernheim a civic leader. He commissioned a statue of Jefferson to stand in front of the courthouse, and in the 1920s bought 13,000 acres of forest land south of the city. "I visualize a natural park," he wrote, "where there will be in profusion all things which gladden the soul." Today Bernheim Forest is the beautiful public park its founder envisioned, as well as the Kentucky State Arboretum and a 12,000-acre research natural area.

It was not Katharine Ordway's choice to be born into wealth, but when a passion for plants sank deep roots in her, she chose to reinvest in the land. In the last thirteen years of her life and through her foundation after she died in 1979, the heiress to the 3M fortune gave at least $64 million to preserve wildlife habitat. Her generosity helped save more than seventy natural areas across North America, including Konza Prairie in Kansas. Ordway money helped The Nature Conservancy create prairie preserves across the Great Plains, and grow into the world's largest, best-known conservation group.

When Henry David Thoreau climbed to the ridgeline of Maine's Mount Katahdin in 1846, he was buffeted by winds and weather, later

Grand Teton National Park, Wyoming, U.S.A. Photo © Antonio Vizcaino/ America Natural

recounting how he was "deep within the hostile ranks of clouds." The scenery, "was vast, Titanic, and such as man never inhabits," he wrote. Long before Thoreau's time, the Penobscot people had venerated the mountain, and Mainers have ever since. Even so, Maine Governor Percival Baxter failed in the 1920s to convince the state legislature to buy Katahdin and the surrounding wilderness from the paper company that owned the land. And so he bought it himself. Over thirty-two years and dozens of transactions he assembled 200,000 acres, and donated them to become Baxter State Park, New England's largest wilderness area.

The original Sevilleta land grant made by the King of Spain in 1819 covered a vast expanse of New Mexico desert and mountains. Nearly 120 years later, General Thomas Campbell bought the land, and the Campbell Farming Corporation operated a cattle ranch there for decades thereafter. In the 1970s, his daughter Elizabeth-Ann Campbell Knapp lobbied the Interior Department to accept the property for a new wildlife refuge. The government eventually accepted, and today the Sevilleta National Wildlife Refuge stretches across 360 square miles of wild country, where bands of pronghorn run free, golden eagles soar on thermals, and the University of New Mexico operates a biological research station. In acreage, the Campbell family's gift was probably the largest of its kind in U.S. history.

I'll conclude these examples with a comparison: Mary Eugenia Wharton was the daughter of a railroad station agent in a small Kentucky town. Her childhood interest in natural history blossomed into a scientific career, and as an adult she chaired the biology department at a small Southern Baptist college, wrote field guides to the state's wildflowers and trees, and was an active conservationist. Even while teaching, writing about, and defending Kentucky's natural heritage, Mary Wharton was also, in her private life, quietly building a sanctuary of her own.

From 1958 to1989, in at least fifteen separate transactions, she purchased land along the Elk Lick Creek outside Lexington. Eventually she assembled a 278-acre preserve called Floracliff and a nonprofit organization to oversee its stewardship. When she died on Thanksgiving Day in 1991, Wharton left an estate of roughly $800,000 to endow the sanctuary. Floracliff is a modest place, hemmed in by a highway and the municipal water company. At Mary Wharton's direction it is not a park, and is closed to the public except for guided tours. It is a true sanctuary, a refuge for wildflowers and warblers, and the occasional river otter who splashes in the creek as it pools and drops toward the Kentucky River.

Conversely, Grand Teton National Park is anything but modest; it offers some of the most spectacular scenery on Earth. The conservationist behind its expansion was born to phenomenal wealth, and spent a lifetime systematically giving it away. And the struggle to create the park was hardly a private affair; it was perhaps the most rancorous, sustained battle in American conservation history, taking more than fifty years to conclude.

That story has more twists and turns than the Snake River as it meanders through the valley called Jackson Hole at the base of Wyoming's Tetons Mountains. The valley was already being marred by tacky development in 1926 when Yellowstone Park Superintendent Horace Albright escorted John D. Rockefeller, Jr. and his wife Abby around the area. Rockefeller asked what might be done, and Albright outlined a plan whereby private lands in the valley might be purchased for a future national park. Rockefeller eventually bought 32,000 acres, but political opposition blocked the land from being added to the original, small Grand Teton National Park designated from federal lands in 1929.

By 1942 the fight had been variously boiling or simmering for decades with the sentiment of local popular opinion shifting back and forth over time. Finally, Rockefeller tired of waiting for the government to accept his gift of land in Jackson Hole, and reluctantly let President Roosevelt know of his decision to sell the property. Whether bluff or pressure tactic, the threat worked and Roosevelt designated Jackson Hole National Monument in 1943, encompassing both mountain and valley lands. A political furor ensued, but prosperity tended to heal old wounds and within a few years even some strident park opponents admitted they had been wrong.

Rockefeller formally deeded his property to the government, and Grand Teton National Park in its modern form was designated in 1950. Can anyone today imagine a better use for this landscape than as a national park?

Mary Wharton and John D. Rockefeller, Jr. led very different lives. She remained unmarried and childless; he fathered six children who became political and business leaders. But they shared a love of wild beauty and a desire to be useful in a way that transcends a brief human lifespan. Surely this also was the motivation also for Percival Baxter when he wrote: "Man is born to die. His works are short lived. Buildings crumble, monuments decay, wealth vanishes, but Katahdin in all its glory forever shall remain the mountain of the people of Maine."

I find these stories inspiring. I hope that as a community of wilderness advocates will come to know them, in the way we know John Muir fighting the Hetch Hetchy Dam, or Bob Marshall founding the Wilderness Society. I hope that we will celebrate the noble tradition of wildlands philanthropy in a way that fosters its resurgence, always stressing that it complements a strong public commitment to conservation.

From southeast Alaska to the Sea of Cortez, from the Great Plains to the hills of Appalachia, from the forests of New England to Caribbean waters and beyond, wildlands philanthropy's potential is limited only by the imaginations of the people who choose to leave a legacy on the land, who spend their energy and their dollars to ensure that wilderness and wildlife will flourish forevermore.

How We Got Started in Chile and Argentina— and Where We Go from Here

Doug Tompkins

Founder, Conservation Land Trust,
Conservación Patagonica

In 1988 I was approached by Rick Klein of Ancient Forests International, a small NGO in northern California, who wanted some funding for the purchase of a small araucaria forest in Chile's 9th region. He had also contacted Alan Weeden of the Weeden Foundation in New York, and my close friend Yvon Chouinard, owner of the Patagonia outdoor clothing company. As it turned out, we all chipped in and the forest of around 1,000 plus acres was purchased, sight unseen. Shortly thereafter I added on to that property; this was my first stab at land conservation. It turned out to be a bellwether for what was to come.

In 1990 I got out of the business world and began looking around for a new place to live, a place far away from cities and with some farming and gardening possibilities. For more than thirty years I had been visiting Chile, and it was one of my first stops in my search for a new home. I met up with Rick Klein and, as we hiked through Alerce forests, he told me about a 16,000-acre farm for sale about 150 kilometers south of Puerto Montt at latitude 42 ... and, that was that. Once the papers were signed, my life took an enormous shift. I moved back and forth between Chile and the States for two years; finally I settled there permanently in early 1993. In late 1993, Kristine McDivitt came to Chile and a year later we got married. She and I have become more than husband and wife, and have formed a true partnership, working interchangeably in our two land conservation trusts, The Conservation Land Trust and Conservación Patagonica (formerly the Patagonia Land Trust).

After the purchase of that first forest ranch, we purchased a very large property of 450,000 acres and formed the Pumalin Project. In the intervening years we added properties to bring the conservation lands close to 800,000 acres, or roughly the size of Yosemite National Park. The process included years of political opposition, many years of working on title disputes, and many more years of building public access. Finally, after many years of waiting, in August of 2005 Chile's President Ricardo Lagos inaugurated the Pumalin Park as an official Chilean Nature Sanctuary, therefore giving the project an institutional blessing by the Chilean state.

During the past fifteen years we have taken on thirteen conservation projects in Chile and Argentina. Two are already national parks: Monte Leon National Park in the province of Santa Cruz, Argentina, and Corcovado National Park in Chile, both coastal parks. In total, we have put nearly 2 million acres of land into some form of permanent conservation.

Currently, Conservación Patagonica, founded by Kris, is working on a new national park, Patagonia National Park in Chilean Patagonia. I am concentrating on a large wetland/savanna project in northeastern Argentina in Corrientes Province. But that does not mean that we are not involved with the each other's work!

Mother and baby guanco. Photo by Kris Tompkins

We work in a variety of very distinct biomes. In Chile we have temperate rainforest, which is far more threatened than tropical forests. Because they have been overshadowed by the glamour of tropical forests, they need extra conservation efforts or will completely disappear. There are between 3% and 5% of original temperate rainforests left in the world. We also work in the Patagonian arid steppe, and subtropical wetland/savanna. Although there are sound arguments for concentrating conservation efforts in hotspots that address conservation priorities (and we applaud the work being done using that criteria), both Kris and I also recognize that conservation work is driven by where you put your own roots down. People come to regard ecological integrity of their home territory as equivalent to the social integrity and the character of the people living there. Wildness and wilderness can and should be everywhere as a core human value.

I believe this from my own experience. Intellectually, I am more certain of it by reading David Ehrenfeld's essay, "Hotspots and the Globalization of Conservation." In short, the world is a huge place despite the claim that it is getting smaller; in fact it has not shrunk a centimeter as far as I know. Erenfeld reminds us that everyone, everywhere, must come to the realization that where they are is where it counts, whether it's a hotspot or not. Our conservation projects are where we

Landscape with laguna. Photo © Antonio Vizcaino/America Natural

live. Fortunately for us, they are glorious places and places we love, and we wish to contribute to their conservation and ecological integrity. I love the tropics, but they are not my favorite place nor are they where my wife and I would like to live. We personally identify with other landscape types more. Now we live half the year in northern Argentina and half the year in southern Chile. Although we are aware of the compromises we make both in our own personal lives and in our work by splitting our lives between two distinct locations, we do have an intense time in both places. We are involved in not just the physical environments, but also the communities where we work, and the local and regional politics. As those working in conservation know, you have to get to know the local scene, meet people, understand the local realities, and of course learn about the local biomes and what conservation strategies would work best in that place. It is perhaps possible to import support from far away, from the "richer" parts of the planet and from those who are willing and proud to help faraway places, but the fine-tuning of strategies and work cannot be imported—they need to be implemented on the ground, day in and day out.

There are arguments for the necessity of outside support in an area that is falling apart ecologically and losing its wildness and wilderness. This is not only financial support. Equally important are fresh eyes, fresh blood, and more objectivity, which give the situation a new perspective. Environmental and conservation history is full of stories of locals trashing their homeland, be it a farm, a town, a county, a region, or a nation. It is extremely rare in the industrial and developed or developing world to find the locals taking care of their ecosystem. Some pre-industrial peoples did. There are indigenous peoples who tried unsuccessfully to defend their territories from the predations of expansionists and imperialist colonizers. They too, however, often diminished the ecological integrity of their own landscapes, though their impact is obviously minimal set against the rapacious work of the industrial economy. Outside perspective can bring new attitudes and new strategies to a local situation. In the places where we work, at times an outsider's perspective is an important and vital catalyst to galvanize a conservation ethic that lies sleeping in a few locals. Often an outsider sparks the imagination, awakens this spirit, and articulates the need for local pride of place. It is almost counter intuitive, but we have seen this happen in each community where we have been involved with projects. Young people are almost always the first ones who take to this newfound conservation ethic. Over time, if one is consistent, persistent, and prepared for the hard work and headaches that serious conservation requires, on-the-ground results can happen.

There is no magic formula capable of being applied universally to land conservation. There are some similar approaches, however, in a larger philosophical sense, like demonstrating the value of the place itself and building local pride, so often missing in degraded or soon-to-be-degraded landscapes. This applies everywhere, but it always takes a fight. There are always entrenched interests somewhere who will suffer the consequences of the curtailment of their enterprises or reduction of their power. Not willingly, they often win out over even well-conducted conservation efforts. If (in the rare event) those opposed to conservation are not present, the area in question most likely does not have a conservation problem to solve!

Alligator having a small meal of a carpincho. Photo provided by Kris Tompkins

What we have learned is that the conservation road is a tough one. We've learned that one will make all kinds of mistakes and nothing makes up for hard work and persistence. Remember though, the opposition also makes mistakes that even the playing field. We have found that every situation is unique, with no formulas or no pat solutions, and generally it takes a long time to get traction and have results. One has to be patient, socially responsible, and sensitive to the local and political realities. But please do not follow the present fashionable thinking of not mixing conservation with environmental activism. Most importantly, we've learned that our conservation goals go hand in hand with our activist work. Looking back over the years, we see that we have accomplished ten times more by mixing these two things. This is not evident at first, and the initial setbacks seem like serious mistakes, but in time and with perseverance you discover that land conservation mixed with activism is far more effective than when these two things are separate (and often in conflict with each other). This will miff some in the land trust movement, I realize, but this is our on-the-ground experience. I think activists who have also had their hands in land conservation would agree. I hope we can encourage our colleagues in land conservation to be more active.

CHAPTER 7

Financing
Wildlands and Wilderness

"Coming Toward"
~

SETH KANTNER

Honorable Mention,
8th World Wilderness Congress Poetry Contest

When all else fails
all else fails
and I'll be alive again
fixing broken things instead
of swallowing plastic smiles.

On the hour when all the ice
melts into mud and beaver tails
I'll walk across water
holding hands for no one.

The sun won't exactly shine
but warmth will come
enough to not build fires
and I will unpack all
this pain, leave it beside a lake.

Twilight is where I'll hide
in the spaces animals appear from
and you—whatever, whomever you are
will walk out of willows
and say "finally!"
And whether you are female,
lizard or my own silence
we are coming toward each other
even now.

The Rise of Natural Capitalism and the New Frontier of Conservation Finance

William Carey Crane, B. Chandler Van Voorhis,
Jerry "Dutch" Van Voorhis
Principals, C2 Invest

For many years our economy has not rewarded the natural assets around us. But as the balance of Earth's resources shrink, amidst a steadily higher level of human consumption and overall growth in world population, technology alone will not supplant extractive resources quickly enough to be beneficial to the health of the planet. Meanwhile, that health is increasingly jeopardized and thrown into a margin of ever-greater risk by the known claims of science that document the dangers of climate change. Suddenly, a new perception is taking hold. This perception is that the frontier of greatest economic opportunity will be for those enterprises that do the following:

- Put more into the earth than they take out
- Engineer products that have no waste
- Reduce carbon emissions whose future costs are steep
- Gain shareholder value through competitive market share growth based on sound environmental practice
- Invest in assets that are lowly correlated to traditional bonds and equities
- Exercise innovative and entrepreneurial responsibility to meet resource shortages
- Embrace the force of natural capitalism over industrial capitalism

The Business of Restoration

The field of restoration economics is a new industry in our economy. The second industrial revolution is being followed by a third industrial revolution: replacing resources extracted from the earth. Technology has been seen as a cure for how to allow an expanding population to survive in a fixed world of resources. But technology cannot keep up with consumption, and both are exerting pressure on the natural resource base of the world. If nature's growing imbalance is to correct itself, nature itself must play a part.

Tree sequestration on scale will have many positive impacts in a restoration economy. The obvious impact is on natural systems themselves. Sequestration will improve flood flow attenuation rates. It restores dirt to soil, contaminated water to pure water, and noxious air to clean air. These impacts are direct. There are also indirect benefits from trees as restoration. They touch the land itself. An acre of ground restored will not only lose its negative opportunity impacts downstream on air, soil, and water; but it will also return an ecosystem to its natural composition: trees will feed the creation of nutrients and restore wildlife habitat essential for nature's own creative process. The indirect benefits go beyond the land also. A restored acre of ground will have a multiplier effect, a positive opportunity cost, in its impact downstream in sustaining better water, soil, and air for space beyond the acres themselves.

Restoration has not been integrated into the economic value system to this point because the natural environment has not been costed. Without a price, it has no value. But as the liabilities of a dwindling supply of land and natural resources and their impacts on the sustainability of our population increase, value and price are being assigned to what has been considered an externality. Restoration economics is ultimately about removing the environment as an externality and making it part of the business structure of the twenty-first century and economy. In that light, we have created the first large-scale restoration company bringing private capital to restore private lands. We call this project company MAP because we have a new compass. MAP is, therefore, a force of natural capitalism—an economic enterprise in which capitalism as we know it will become capitalism as it will be known. Through a project such as MAP, the environment will come to have a capitalistic edge, and capitalism will have a new profit line that is connected to purpose. It is probably less important to use Corporate Social Responsibility (CSR) as a process, than to put the process to a purpose that is larger than the corporate window itself.

That which gives urgency to restoration economics in a carbon-constrained world is environmental danger. In the past, problems in the environment could be cited and repaired with known past liability and cost. Climate change reverses course. It is an unknown future liability with unknown costs, and it cannot be repaired. So it must be mitigated and not just moderated, and tree sequestration is a major means of not only moderating the impact of carbon emissions but actually mitigating the creation of those emissions. MAP is a mitigation strategy for climate change on a large enough scale to achieve impact through a combination of market forces that create a sustainable investment process.

To date, our means of dealing with conservation problems have been outside the markets. We have relied on charity as a "thousand acts of random kindness," or the force and foresight of governmental regulation. The first has great impact, but in limited duration. The second is not responsive to the creation of a restored economy, only the lessening of damages to one not working well with the environment. If, however, the market mechanisms can be part of underwriting landscape change, the future is brighter. MAP takes this new road, and takes it boldly.

MAP will produce significant renewable energy supply and have a meaningful positive effect on climate change through the sequestration of carbon dioxide effects, which will ultimately benefit the energy sector of our nation.

MAP is a means of entering into the world of energy agriculture outputs. In addition to MAP's supply of high-quality carbon credits, MAP will also supply another asset that directly benefits the energy industry: biomass. By co-firing biomass with coal, industrial plants can reduce their emissions and qualify for a Production Tax Credit based on the percent of kilowatt hours (Kwhs) produced from biomass. For utilities whose generational capacity relies heavily on coal firing, biomass is a key energy product strategy that also satisfies carbon compliance environmental objectives. MAP is capable of supplying annual biomass within ten years of its inception. This steady supply allows utilities to develop a consistent plan for emissions reductions.

The Role of Technology

Technology and new forest techniques are incorporated in MAP's ecological and economic goals. They are harmoniously joined through a new practice of using high-grade genetics with tree planting geometry that is rooted in sound forest development as well as strong economic forest production. There are some tracts with experience in this technology. MAP will be the first large-scale employment of what has become known successfully as interplanting technology.

Traditionally, many of the values and income of a forest are not captured because they are left to nature to manage and the time frame is elongated. A proper forest management plan enables the realization of values and income streams before they decompose. MAP employs a cottonwood interplanting technique that mimics nature and enables time and space to be compressed so that the layers of values materialize faster. This is critical in delivering carbon benefits that help the biological clock fit the market window while expediting the delivery of a forest with all its conservation benefits.

Cottonwood acts as a nursery crop. Because it is a fast growing tree that reaches heights of eight to twelve feet in one year, it provides structure for the hardwood and increases the hardwood's survivability. But more importantly, cottonwoods create a canopy faster. It is this canopy that speeds up the biological activity on the ground floor. Once the cottonwoods reach 100 feet tall by year ten, they need to be thinned in order to bring light to the forest floor, thus enabling the hardwood to grow more quickly. It is this act of realizing income that produces appreciation of the hardwood by speeding up its growth and producing a straighter more economically valuable tree. Tree production should be aimed at delivering a straighter tree. The fact that the trees are tightly packed together forces them to grow up and not branch out as much.

MAP's cottonwood interplanting arrangement enables a natural control mechanism for mitigating disease and insect infestation, weed control, and achieving increased survival of both cuttings and seedlings. Furthermore, the interplanting arrangement accelerates establishment and growth rates of hardmast producing hardwood species, as well as light seeded hardwood. MAP's interplanting arrangement accelerates the typical succession and maturation cycle from 150-plus years to an impressive thirty- to seventy-year succession and maturation cycle.

MAP achieves its impressive alternative silvicultural strategy through process enhancements of selected task categories in the standard industry reforestation implementation model. These combined process enhancements include:

- Compressed initial planting cycle from fifty-one months to twenty-seven months
- Accelerated tree growth
- Increased tree quality
- Improved tree harvesting cycles
- Expedited business processes
- Mitigated project risks

Why the Delta?

Historically, the Lower Mississippi River Valley (LMRV) existed as the North American equivalent of the Central and South American tropical rainforests. Previously a 25 million acre temperate rainforest, the LMRV region still provides exceptional tree growth potential as well as important plant and animal biological diversity. Habitat requirements for a host of neotropical migratory songbirds directly link the biological diversity values of the LMRV's rainforest to the Central and South American rainforests. These migratory songbirds depend on the

Central and South American rainforests for their winter habitat needs, and they return to the LMRV's temperate rainforest for their spring and summer needs as well as to raise their young. Unless both ecosystems are conserved, numerous species of neotropical songbirds will not be sustained.

Destruction of the LMRV forest came primarily at the hands of mechanized agriculture following World War II, with the most dramatic destruction period being the mid 1960s through the mid 1970s due to extreme increases in soybean prices. In recent years, new reforestation processes have begun to take shape through a number of government programs. Unfortunately, the private sector has made no concerted efforts to build upon the initial government activities. Without private efforts, the vital restoration of this temperate rainforest, on a scale necessary to return biological diversity value to the region, will fail. MAP is ideally positioned to assist with this vital reforestation effort.

With the development of the new cottonwood-hardwood interplanting technology, an economically viable reforestation technique is emerging. The technique will enable investors and landowners to achieve both a more timely return on their investment while achieving the restoration of the full range of temperate rain forest natural resource values. MAP has successfully established the largest cottonwood nursery in the region and has included in it the most advanced genetic cottonwood clones, capable of growth rates that are superior to all others. MAP staff and advisors have in depth knowledge of the Lower Mississippi River Valley as well as highly credible technical expertise, and are ready to begin the reforestation process. Close working ties with governmental agency personnel in the region who have been involved in government reforestation activities have been established, and through partnership activities with these governmental personnel and their programs MAP is poised to begin implementation of a large-scale reforestation effort of national significance.

A New Business Paradigm

Overview of MAP

MAP is a conservation project, finance company founded by C2I's principals, Carey Crane and Chandler Van Voorhis. Whereas most projects are created to oversee a depleting asset by extracting resources from the earth, MAP was designed to give the world a permanent asset of greatly increased value through reforestation and carbon sequestration.

To achieve this investment opportunity, three partners (the landowner, government, and investor) must interface in a classic public-private partnership. MAP's initial reforestation focus is primarily the lower Mississippi River Delta including acreage in Arkansas, Louisiana, and Mississippi. This effort is designed to integrate MAP efficiencies and incentives into an existing farm subsidy program administered by the United States Department of Agriculture's Farm Service Agency (USDA FSA), of which certain landowners are enrolled. Investor returns from MAP are targeted to meet or exceed returns currently reported by Timber Investment Management Organizations (TIMOs).

The Challenge to Conservation

As land is changing hands and possibly use the concentration of wealth in the world of conservation money becomes ever so slightly diluted, making it harder for conservation to rise to pressing needs. In the world of forestry, this has never been more evident. As Wall Street's appetite for an alternative class that produces low correlations to the bond and equity markets heightened after 9/11, the vertically integrated forestry company is flattening out and selling its high multiple assets (land) to pay down its debt and improve its short-term stock price. Conservation has been forced to play on the retail side of transactions. This comes at a high cost to conservation.

The Role of Private Capital

Since most of conservation has been funded through philanthropic or governmental money, private capital in conservation represents a new frontier. Bringing private capital into conservation forces, the conservation mindset to look to returns both economic and environmental. We call this the blended return. As conservation becomes more centered in economics and economics become more focused on purpose, the blended return becomes a powerful and dynamic proposition, enabling a wider net of available dollars available to create exponential impact.

A sustainable investment is a good thing. But a sustainable investment that is hooked to a sustainable investment process becomes a force for exponential change. MAP, by design, has been engineered to create a sustainable investment process through large-scale replication. We believe these are fundamental design elements of natural capitalism.

MAP's Vision and Goals

The creation of a forest produces layers of value. Some we can price today, others we cannot.

In order to capture all the values, MAP employs a concept called ACRE (patent pending), or Advanced Carbon Restored Ecosystem. The ACRE is a series of ecological assets that represent all the values and attributes associated with one acre of restored property. Much in the same way that Renewable Energy Certificates (RECs) capture all the environmental attributes associated with one megawatt hour of power generation, ACRE's value stream is rooted in a common, traditional denominator: one acre of property. ACRE is an appreciating asset that produces valuation and uneven revenue streams. Investors will purchase these assets for a period of seventy years through a lease of rights rather than fee simple ownership of land. The rights of the ACRE translate to an equivalent number of acres of land the investor's money will touch.

MAP has five goals. The first is to bring investors to conservation on a large scale for the first time and use market forces rather than charity or regulation to achieve this goal. The second is to plant, as a first initiative, 500,000 acres of trees. MAP anticipates this will be accomplished in five tranches of 100,000 acre Conservation Reserve Enhancement Programs (CREP). The third is to use the production of those trees to achieve significant energy outputs in timber and the emerging ecological markets of carbon and biomass through optimization of these and other ecological assets. The fourth goal is that each of these three things will result in a mature, mixed hardwood forest of the highest standard of conservation practice. The fifth is to create a replicating model.

The Design Principles of MAP

MAP rests on five integral points, each of which support the other four.

Land

Many good things have been done on the land. But few have been done on enough of it to change the value of the land. MAP believes in scale. If enough land is dedicated to a core purpose, the seismic effect of scale can make a difference in the landscape as a whole, both the acres touched and all the ecological elements below, around, and above them.

Lease

Land is affected by degrees. The price of land is a constraint to making a difference in the land. Our system requires property rights as the vehicle for control of choices on how to treat lands we own. But what if we don't own the land? What if the price of ownership is offset through the lease of land assets? Leasing land gives effective control of the same rights. Moreover, the dollar goes much further. One acre of owned land may be five acres of leased land. The difference can be greater to the landscape if the cost of ownership is not part of an environmental transaction that needs to occur on scale.

Leverage

MAP gets leverage through leasing, and impact through scale. By linking its approach to environmental change and to existing sources of capital outside land that is leased to achieve scale, MAP can further leverage its change base. In particular, government subsidy programs can help finance long-term project costs more effectively by being leveraged monies rather than direct, terminal outlays. MAP has designed how to combine this leverage with the lease and land scale principles.

Layers

Once MAP secures land on scale, it is able to increase the return of the land asset by creating layers of value in the asset. In its reforestation efforts, MAP will build on the lease and leverage values by seeking present and future economic returns not only on timber, in this case, but its derivative assets. The full ecology will be part of the marketplace. First, good genetics will grow good trees. The planting areas will have high tree growth rates generally. The cottonwood as a nurse crop will store carbon faster. Biomass from the cottonwood will be a diversification strategy for pulp. The ecological benefits of a fast growing but healthy forest will be marketed. Presently these include: nutrients, water credits, carbon, and conservation banking credits. A greater return will be felt on every acre as a result. In carbon, for example, the ability to generate credits will not only be an offset but also a reduction of carbon dioxides that cleans the environment while selling from it.

Legacy

The outcome of this economic engineering that respects rather than violates nature will be a conservation legacy. Earth will have its resources drawn from it to make a living, while restoring them to the earth because of that living. This reverses the long-standing industrial tendency to exploit Earth for gain. The exploitations return gain to Earth, leaving nature better off than it was found.

Over thirty months of engineering from points of derivative and financial structures, severing of ecological rights, land use, conservation easements and forestry has occurred to bring each of these five main points into sync to a financially ready investment grade product.

Three Layers of Return and the Legacy

The economic returns of MAP are broken down into layers. The first layer is rental income. This layer acts like a bond delivering absolute returns. The second layer is comprised of timber itself. This layer coupled with the first layer creates returns that mirror equities. The final layer of return is made up of ecological assets such as carbon and nutrients. Using modest pricing assumptions, when coupling this layer to the previous two, the returns look more like a venture capital investment. In essence, within the ecosystem is a spectrum of returns that spans bonds, equities, and venture capital.

Through the public-private partnership, the sustainable investment process, the design of replication, and the five integral points previously listed, MAP has engineered a financial product that can be replicated. Using private capital to deliver a blended return to landowners, the government, and private investors, a sustainable investment process will successfully underwrite the financial span a sustainable investment will require in the future to be an effective new paradigm while creating a conservation legacy. Doing well by doing good!

The Global Environmental Facility's Commitment to Wilderness Areas

Len Good

Chief Executive Officer and Chairman,
Global Environmental Facility

In every part of the world—with wars, storms, floods, droughts, exploitation—we see our natural and cultural heritage undermined by human and natural impacts. The importance of safeguarding wilderness for the benefit of life on Earth becomes increasingly self-evident. Our wilderness areas are some of the largest and most important providers of services fundamental to maintaining our planet and our global population's long-term health and stability.

Now more than ever, the world must come together to ensure that future generations are not deprived of the invaluable legacy that the global network of wilderness areas represents.

Our movement continues to grow in importance and stature. The people of Alaska have clearly shown their commitment to preserving and protecting wilderness and recognizing that to do so we must pay attention to human needs.

The modern conservation movement began in the United States with Yellowstone and Yosemite National Parks. The U.S. is a country that has been a strong supporter of wilderness conservation through the support of the World Conservation Union (IUCN), and the largest contributor to the Global Environmental Facility (GEF). The Native peoples of Alaska, like many indigenous cultures around the world, hold tight to two important values: 1. Pride and respect for the land; and 2. A sense of stewardship for the land. It is these beliefs that the World Wilderness Congress and the GEF seek to reinforce and uphold.

The GEF recognizes the historic role that wilderness has played in human civilization and shares the Congress' belief that spirit and culture are part of science, politics, economics, and education. The World Wilderness Congress has also recognized the important role of our younger generations in shaping a future for wilderness, by supporting youth- and young-professional-related activities. On behalf of the GEF, I truly welcome these efforts.

The GEF can trace its origins in part to the 4th Wilderness Congress held in Colorado in 1987, where a working session hosted by The Wild Foundation conceptualized a study by the World Resources Institute. The study called for an international facility to help reverse the rapid environmental degradation faced by developing nations. This study led to essentially what the GEF is today. So I hope the Congress is proud of that contribution and that the GEF has lived up to the expectations.

Over the last fourteen years, the GEF has grown to become the catalyst for actions to improve the global environment in partnership with

176 member governments, 700 NGOs, the private sector, and indigenous groups. It mobilizes international cooperation, linking local and global challenges, helping to move the world towards a sustainable future.

Let me give a few numbers that reflect our support and enthusiasm for biodiversity conservation and sustainable use:

- More than $2 billion to about 700 projects in 155 developing countries
- The single largest block of funding worldwide
- Leveraging an additional $4.6 billion in co-financing from project partners
- Direct and indirect support to wilderness areas
- Almost $1 billion in grants supplemented by more than $1.8 billion in co-financing

Alaska is a state known for its expansive and striking wilderness areas, as well as its wilderness values and ideals. It is a place with a long history of interaction between humans and their environment—people have made this land their home for thousands of years. Alaska embodies each aspect of the wilderness concept—ecological, cultural, and spiritual. Because of its great size, Alaska also contains many different types of critical ecosystems, from Arctic tundra to coastal temperate rainforests. And, as is increasingly being demonstrated, these ecosystems serve as barometers of global climate shift, reflecting changes before they are seen in many other parts of the world. And they are increasingly threatened.

Alaska, like most places in the world, faces difficult choices in finding ways to maintain intact ecosystems while ensuring that the needs of people are met, both in the short and long term. But Alaska, with an average population density of a little over one person per square mile, might seem to have more room for accommodation than many other parts of the world. GEF's role is to find ways to meet these challenges, and ensure that the value of wilderness is not eroded, even in areas that have less room for accommodation than Alaska.

We do so, noting:

- Wilderness areas bring significant benefits both to the communities within them and to the larger global community
- Indigenous knowledge contained therein is vital to conserving biodiversity and managing wilderness areas
- The full potential of wilderness areas is realized only when their immediate geographic, economic, and social contexts are taken into account

Specific examples of projects in which we are involved:

- The Kamchatka Peninsula, Russia
- Forests and rivers of Southeast Asia and central Africa
- The deserts of Jordan, the Gobi desert of China, the Kalahari
- Peru, Bolivia, and Brazil

In Brazil, the GEF supports the Amazon Region Protected Areas Program (ARPA), which currently has approximately 12 million hectares of tropical forest with a goal to bring a total of 37 million hectares under strict protection, the biggest joint initiative for the conservation of tropical forests in history.

And let me mention the Baviaanskloof Wilderness Area in South Africa. At the 7th Wilderness Congress (2001) held in Port Elizabeth, South Africa, the GEF announced that it would provide $1 million in support for the Baviaanskloof initiative in the Cape Floristic region, one of the most biologically diverse wilderness areas today. I am pleased to note that the GEF has continued to collaborate with the government of the people of South Africa in the conservation of the Cape Floristic region. And I am also pleased to note that the South African government announced its intentions to designate Baviaanskloof as the largest wilderness area in the world.

Small Island Developing States (SIDS) also receive support from the GEF. SIDS are repositories of some of the world's most diverse and unique ecosystems and species. Since its inception, the GEF has allocated approximately $180 million for 115 biodiversity projects in SIDS.

Clearly there are wilderness areas around the world that need to be managed and expanded. I am impressed with the initiative taken at each World Wilderness Congress, especially The Wild Planet Project, whose objective is wilderness management and expansion. The GEF will collaborate in this major effort, and hopes to see its initial emphasis in South America spread to Asia and Africa.

Looking ahead, the GEF's strategic focus in the coming years will be on:

- Catalyzing the sustainability of systems of protected areas across the globe

- Continuing to support the larger production landscape as it contributes indirectly to the integrity of protected areas and takes into account the needs of local communities and indigenous peoples

- Finally putting additional emphasis on marine protected areas, which typically are underrepresented across the globe, thereby contributing to the growing momentum of establishing marine wilderness areas worldwide

In closing, let me say the conservation of our wilderness areas and natural heritage will require the continued commitment of all of us, working individually and collectively to realize our mutually shared goals. I want to assure you that the GEF stands ready to continue working with you towards achieving those objectives.

Chapter 8

Global Issues

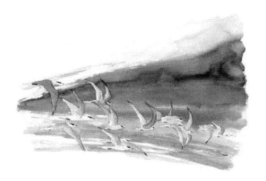

"Birders"

~

RICHARD CHIAPPONE

Honorable Mention,
8th World Wilderness Congress Poetry Contest

You start with an Instamatic
and crumbs in the back yard
and work your way to optics
worth more than your car,
trips to the tropics.

Now you watch your life list grow
and shop the Net for gourmet suet.
If your office had a window
you'd lose the day peering through it
at chickadees and sparrows.

You know the names of birds so dull
their mothers can't tell them apart,
and can replicate the call
of two kinds of kinglet.
But the fascination is not mutual.

Hawks could train their perfect eyes
on your nest—yet don't.
And only when his feeder's seedless
does that nuthatch note
that you even exist.

The fully feathered truth is
that we the plumage deficient,
the flightless and virtually songless
apparently just aren't
all that interesting ourselves.

The Millennium Ecosystem Assessment
A Framework for Wilderness Stewardship in a Directionally Changing World

F. Stuart Chapin, III
Professor of Ecology at the Department of Biology and
Wildlife Institute for Arctic Biology, University of Alaska

The world is undergoing rapid directional changes in most of the factors that control the properties of natural and managed landscapes. In the last fifty years, humans have changed ecosystems more rapidly and extensively than at any comparable period of human history, with even more rapid and extensive changes projected for the next half century and beyond (Millennium Ecosystem Assessment, 2005). For example, human activities have substantially altered climate, atmospheric chemistry, biodiversity, land-cover, the use of biological productivity, and the cycling of nitrogen (N) at global scales. In addition, as the human population continues to grow, there will be both increasing pressures for use or conversion of current wilderness lands as well as increased need for the ecological and cultural values provided by wilderness to society at large. Given these changes in environmental, biotic, and socioeconomic controls, the future state of wilderness is certain to differ from what it has been in the past. Attempts to preserve wilderness simply by excluding management are therefore unlikely to sustain its essential values.

The Millennium Ecosystem Assessment (MEA) was initiated by the United Nations to assess the consequences of ecosystem change for human well-being and to develop the scientific basis for actions needed to enhance the conservation and sustainable use of ecosystems by society (Millennium Ecosystem Assessment, 2005). Over a four-year period, about 1,400 experts from around the world worked to address these goals. The MEA provides a useful framework to understand the causes and consequences of wilderness change and the important societal role that wilderness can play in an increasingly human-dominated world.

Wilderness is at one end of a spectrum of intensity of human interactions with ecosystems. The MEA identified several categories of ecosystem services (i.e., the benefits that society derives from ecosystems) that all ecosystems provide. Supporting services are the underlying ecosystem and population processes that are part of the essential fabric of all ecosystems. These include the cycling of water, carbon, and nitrogen, and the birth, death, and migration of organisms that determine local patterns of biodiversity. Provisioning services are the goods that people harvest from ecosystems, such as food, fiber, wood, natural products, and water. Although provisioning services are harvested most intensively from agricultural ecosystems, the production of berries, medicinal herbs, game animals, and water in wilderness are important to people who live or visit there. Ecosystems also provide regulating services that provide benefits to society well beyond the boundaries of a wilderness area, such as climate regulation, disturbance regulation, air and water purification, and disease regulation. As the regulating services of managed ecosystems become degraded, the delivery of these services by wilderness becomes increasingly important. Ecosystems also produce cultural services that provide a sense of place and identity, aesthetic or spiritual benefits, and opportunities for recreation and tourism. Although all ecosystems provide this spectrum of services, wilderness tends to be particularly important in providing regulating and cultural services that are essential for long-term well-being.

A fundamental assumption of the MEA is that all ecosystems, including wilderness, are best understood as social-ecological systems whose properties are shaped by interactions between people and the land or sea. Some human interactions with wilderness are diffuse, such as the impacts of anthropogenic pollutants on remote lands or the existence value that many nonresidents derive from simply knowing that wilderness exists. However, people inhabit most wilderness areas in their natural state. The Eurocentric concept of wilderness as an area where "man himself is a visitor who does not remain (the Wilderness Act)" has rarely characterized remote regions that we think of as wilderness (Watson, et al., 2003). In Alaska, for example, people have inhabited remote areas for at least 10,000 years and have depended on wilderness as a source of subsistence resources and as a place to live. Although the tools by which Alaska Natives interact with the land have changed to include rifles, snow machines, and motor boats, which in turn requires integration into a mixed cash-subsistence economy, there remains a strong cultural and economic dependence on the land (Chapin, et al., 2004).

Given that wilderness areas will change, and the nature of these changes will be strongly influenced social-ecological interactions, how can we construct a framework for wilderness stewardship that embraces change as an process that brings both opportunities and challenges rather than simply attempting to prevent change (Gunderson and Holling, 2002; Berkes, et al., 2003)? Based on the experience of the MEA, some general rules emerge that may prove useful (Chapin, et al., 2004; Millennium Ecosystem Assessment, 2005).

Given that change is a natural feature of social-ecological systems, it is often beneficial to foster modest changes rather than preventing those changes that make catastrophic events more likely. Fire suppression, for example, reduces the probability of wildfire in the short term but increases the probability of future larger fires. Crises or other large changes that do occur can be treated as opportunities to think outside the box for novel solutions, to address future needs. Regardless of what happens, it is important to learn from change, because it is virtually certain that changes will continue to occur.

Sustaining diversity provides more options to respond effectively to changes that occur. For example, maintaining large management units with a wide range of ecological and topographic diversity provides opportunities for organisms to migrate in response to future climate changes rather than being trapped in a local preserve that becomes gradually less suitable as habitat (Elmqvist, et al., 2003). Similarly, fostering cultural diversity in which people with different cultural ties to the land (e.g., subsistence users and backpackers) provides opportunities to interact with the land in different, but equally appropriate, ways.

Planning for the long-term integrity of wilderness in the face of certain changes in climate, culture, and economy is a serious challenge, but represents an opportunity to think creatively about the deepest values that underlie the human need to be a part of wilderness in an enduring fashion. Resident and nonresident users of wilderness must work together to define the limits to acceptable change and the attributes of wilderness that are most important to sustain.

References

Berkes F., Colding J. & Folke C., editors. 2003. Navigating Social-Ecological Systems: Building Resilience for Complexity and Change. Cambridge University Press, Cambridge.

Chapin F.S. III, Henry L. & DeWilde L. 2004. Wilderness in a changing Alaska: Managing for resilience. International Journal of Wilderness 10:9–13.

Elmqvist T., Folke C., Nyström M., Peterson G., Bengtsson J., Walker B. & Norberg J. 2003. Response diversity, ecosystem change, and resilience. Frontiers in Ecology and the Environment 1:488–494.

Gunderson L.H. and Holling C.S., editors. 2002. Panarchy: Understanding Transformations in Human and Natural Systems. Island Press, Washington, D.C.

Millennium Ecosystem Assessment. 2005. Ecosystems and Human Well-being: Synthesis. Island Press, Washington, D.C.

Watson A., Alessa L. & Glaspell B. 2003. The relationship between traditional ecological knowledge, evolving cultures, and wilderness protection in the circumpolar north. Conservation Ecology 8:http://www.consecol.org/vol8/iss1/art2.

The Global Commons

The Honorable Walter J. Hickel
Former Governor, State of Alaksa, Former U.S. Sectetary of Interior

As we marched on the cobblestone streets of Stockholm, the local citizens stared in amazement. We were a rag-tag group, hundreds of mostly young people from around the world who had come to the first Earth Summit in 1972. We were determined to focus the world's attention on the environmental crisis and the need for action. Leading our parade was an enormous whale, made from chicken wire and polyethylene, stretched over an old hippie bus. As it rumbled down the streets, the creature broadcast the cries of whales recorded deep beneath the sea. Our goal was a ten-year moratorium on commercial whaling.

I attended that historic meeting as one of six World Observers who, just two years before, placed all eight species of great whales on the U.S. endangered species list. But the delegates of that summit were deadlocked. Something bold and unconventional was needed. With the help of volunteer activists, we built the whale and began the march. As we approached the Old Parliament Building, hundreds of Stockholm police locked arms to block our progress. They braced for a clash. But there was none. We smiled and handed them orange blossoms.

The media got the message. Front-page headlines and photos of that march filled the newsstands. The next day, the resources committee passed the whaling resolution 51–3. The logjam was broken and the following day that same committee passed twenty-six more. That whale walk was bold and unconventional, and it worked.

And the world today, more than ever, needs bold and unconventional ideas. Fortunately, most whale species have recovered from the brink of extinction. Today, our target is the human race. Twenty years after Stockholm, at the 1992 Earth Summit in Rio, I urged that global forum not to forget people. I called for a commitment to the total environmental equation: people, people's needs, and nature. There will be no cure for pollution until we find the answer to poverty. We needed a turning point where the protestors of progress became the protestors of poverty.

Indeed, the Rio Earth Summit was a turning point. In their report, titled "Agenda 21," government and NGO leaders expanded, broadened, and deepened the world's understanding of the environment. I describe that turning point this way: They recognized that every creature and every child God placed on Earth has a purpose. They acknowledged that any man, woman, or child who is cold, hungry, or unemployed is in an ugly environment. They declared that when a child dies of hunger in her parents' arms, the seeds of terrorism have been planted.

In my eighty-six years on this planet, I have learned that to succeed at anything, you have to believe. Victory will be won or lost within us. When the world awakens to the life and death urgency of the issues we are facing today, money will be no obstacle. We must mobilize the commitment of people.

Every one of us can make a difference. Think of what 6 billion hearts, 6 billion minds, and 6 billion pairs of hands working together can accomplish! Every generation faces a challenge, and this one is ours.

Many parts of the world continue to struggle with old-fashioned and destructive approaches to economic and social betterment. Countries and cultures are locked in mortal combat between worn out ideologies, militant religions, and partisan politics. Alaska's system is unique because we are governed by a constitutional democracy, but we live from the resources of both land and sea that are owned by all. We call it "the commons."

Other than native lands, which are communal in their own way, there are almost no private lands in Alaska. We survive and subsist from the commons. We make our livings from the commons. We fund government from the commons. And we know very well that the commons is both valuable and vulnerable. It is easily exploited by those who come, take, and leave nothing, or by those who lock it up and do nothing.

Our goal, which is clearly stated in our constitution, is that the riches of the commons must be utilized for the good of all our people. Not just for the benefit of the multinational corporations, politicians, or those we used to call robber barons.

The commons is not unique to the North. On all continents, vast reaches of commonly owned lands and waters are owned by states and nations, governments and monarchs. And beyond that, the vast majority of the commons is the unclaimed land and sea, the air, the water and even space, owned by no one and therefore owned by all.

Most people from the cultures of the West think of Earth as privately owned, but it's not. At least 84% of the world's surface is commons. Our Institute of the North, based in Anchorage, is developing a world map of the commons. We have a rough concept of what it looks like in the northern hemisphere. That map can be found in my book *Crisis in the Commons: the Alaska Solution.*

But so far we have been unable to unravel most of the ownership patterns in the southern hemisphere. Thanks to Vance Martin and Josh Brann of the Global Environment Facility (GEF), who have spent many hours working on this problem. However, the research librarians at the World Bank told Josh that such data does not exist. So did the senior scientist at the Land Tenure Center at the University of Wisconsin, Madison.

In any case, Josh, who also happens to be an Alaskan, continued to work on the question and his research indicates that all nations in Africa, with the exception of South Africa, are more than 50% commons. However, he could find no statistics for New Zealand. And he was unable to determine the status of most of South America. Therefore I ask your help, especially those of you from the southern hemisphere. Please help us complete this map.

Why is this important? Because the commons is a key to the future of the human race. It is one of the missing links in our plan for world survival. The commons is a repository of God-given resources, from clean air and water to minerals and energy. And it could be a source of funding for the United Nations' ambitious Millennium Development Goals.

For years, academics have studied "The Tragedy of the Commons." They have catalogued exploitation, overcrowding, and overuse. Now, it is time to address the opportunity of the commons. Let me explain. ...

The ocean commons provides much of the protein the world consumes. No one owns the oceans beyond the 200-mile limit. No one owns the distant ocean floor and the resources it contains, which are immense. Don't fall for the prevailing wisdom that the world is running out of resources. All we've done is slide around on the skin of the apple. We haven't even thought about the core. When we finally discover the value of the oceans and their subsurface wealth, the onshore lands could be reserved mainly for living and parks. I don't fear that we will run out of resources. I might be worried if this were the only planet we could see.

It's time that the United Nations establishes a Global Commission on the Commons. As the first order of business, such a commission should establish rents and royalties for those who harvest the wealth of the global commons. These proceeds should be dedicated to advancing the Millennium Development Goals.

In recent years, many people have been alienated by the tactics and the abuses of the world's multinational corporations. Having lived in Alaska when it was an exploited colony, I understand their frustration. Some corporations fly no flag but their own. Their only allegiance is to the bottom line. They have been infected with a virus, a virus of greed. But the game is up. They know it, and we know it, and it's time for a change.

Make no mistake; we need the expertise of these companies. There is no wealth without production. So let's invite these companies to be part of the answer. Those who generate wealth from the commons must help the common wealth. In Alaska we call this approach the "owner state." We use our public resources for the benefit of the people. We are no longer poor people living on rich land.

Nations looking for a new approach might take a look. Alaska's owner state is a giant step beyond both failed communism and unbridled capitalism. Ours is an unusual combination; a constitutional democracy, a free enterprise economy, and public ownership of resources.

Russian economists have expressed interest in this model. And I believe that it will work well in Africa, because it works best in cultures with a tradition of community. Most of all, let your minds imagine how our approach to the commons, set in the context of a constitutional democracy, can break the endless cycle of poverty that has plagued mankind.

When we learn to use the world's resources productively for the benefit of all, not for just the selfish or corrupt, the human race can stand together, arm in arm, on the threshold of wealth and social advancement we can now only imagine. We will save and protect the most beautiful lands and wilderness for enjoyment of all, and there will be no legitimate reason for poverty.

CHAPTER 9

Wild Places

"How to Begin to Know this Place"

~

NANCY LORD

Honorable Mention,
8th World Wilderness Congress Poetry Contest

After the alarm clock you meant not to set nonetheless peals
open the day, lie on your narrow cot under the eave and listen.
Name crow, fox, sparrow, magpie, distant gull. Name the crow in
Dena'ina: gguguyni.

Walk to the Cliffside and look under your feet: spruce needle and
cone, felted moss. Inch to the edge, stare down on fast-flowing
channel, bared mussel beds. Pick one feather frond of yarrow.
Bite it, that pepper taste.

Consider the mountains. New snow softens the contours, divides from pewter sky. Lift your face to a fine prickling drizzle.

Lay a hand on old bark. Let your fingers find the furrows, read like Braille the bored beetle holes.

Breathe.

Wilderness across Borders

Trevor Sandwith

Director, Cape Action Plan for the Environment

The high mountain wilderness of the Maloti-Drakensberg Mountain Range straddles the international border between South Africa and Lesotho. Far from the centers of commerce and industry, the mountain ecosystem is not only a refuge for biodiversity and cultural resources of outstanding universal value to humanity, it is also the source of the water on which the people of the subregion depend. For many years, it has been our dream to secure these values by establishing an international conservation area. The pioneering vision of Bill Bainbridge and Bore Motsamai kept this dream alive, even when the two countries were not on speaking terms.

The dream is of a vast transboundary wilderness that not only conserves biodiversity and cultural resources and contributes to peace and reconciliation in the subregion, but also is a source of sustainable social and economic transformation and development.

Transboundary Conservation Initiatives

If one examines the global distribution of protected areas, it becomes clear that protected areas are located disproportionately at or near the borders of countries. This is most likely the result of these areas being less significant for economic exploitation and providing greater opportunity for protected area establishment.

Since 1932, when the world's first officially recognized transboundary protected area was established between the U.S. and Canada (Waterton–Glacier International Peace Park, Figure 1), there has been a

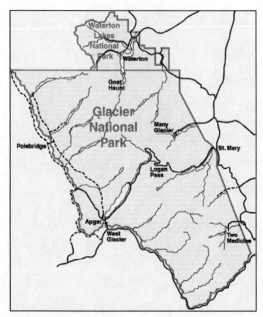

Figure 1 Waterton, Glacier International Peace Park

rapid growth. Dorothy Zbicz's work indicated the rapid rise of internationally adjacent protected areas from fifty-nine areas in 1988 to 169 areas in 2001.

A complete review of the current situation, undertaken in early 2005 by Charles Besancon, recognized at least 188 situations. We know that there are many more situations where transboundary conservation initiatives underpin ecoregional, hotspots, and wilderness programs, and so this growth will continue.

The principal driver of transboundary conservation initiatives is a focus on biodiversity and wilderness protection. The biological rationale is that while the protected area estate has performed relatively well in securing representative samples of biodiversity pattern (distribution of species, communities and ecosystems), it remains inadequate to conserve the ecosystem processes that will secure persistence either of the protected areas or of biodiversity in the wider landscape.

Furthermore, the apparent paradigm shift reflected at the World Parks Congress is that conservation, itself embedded within the socio-cultural milieu of the place, must deliver socioeconomic benefits to society, beyond the borders of protected areas.

Multiple-agency, landscape level approaches, the ecosystem approach, can contribute to the resolution of this problem. The cooperation that is required to mobilize effective implementation also engenders understanding among adjacent countries and management agencies, and in so doing contributes to the maintenance of friendly relations, and even peace.

A number of pioneering studies have helped to characterize and put transboundary conservation on the map. Generally drawing on examples of initiatives and implementation around the world, IUCN-World Conservation Union and the World Commission on Protected Areas in the capable hands of Jim Thorsell and Larry Hamilton put together initial guidance. This has been followed up by substantial useful work by the EuroParc Federation, the Biodiversity Support Program, and the German Foundation for Development. It culminated in the publication of *Best Practice Guidelines* by the IUCN Task Force on Transboundary Protected Areas. Since the World Parks Congress in 2003, this publication has led to ongoing work to characterize and understand the dynamics of transboundary conservation programs.

An important part of this work has been the description and development of four main types of transboundary conservation initiatives. The first of these is the transboundary protected area, where essentially there are two protected areas that conform to the IUCN definition of protected areas (i.e., any of the six protected areas categories). A transboundary protected area can be defined as:

> "An area of land and/or sea that straddles once or more boundaries between states or sub-national units such as provinces or regions, autonomous areas and/or regions beyond the limits of national sovereignty or jurisdiction, whose constituent parts are especially dedicated to the protection and maintenance of biological diversity, and of natural and associated cultural resources, and managed cooperatively through legal or other effective means."

An example of a transboundary protected area is the area between France and Spain in the Pyrenees Mountains. The Pyrenees National Park in France is adjacent to the Mont Perdu National Park in Spain (Figure 2). Of interest here is how the two countries have committed to these national parks being incorporated into a single World Heritage

Site. Furthermore, they form part of a complex system of protected areas of different categories and the collective mosaic is recognized as a Biosphere Reserve.

This leads us naturally to consider situations where not all of the component pieces of the transboundary conservation mosaic are protected areas, but where intervening areas are included, in which other uses are permitted. Leo Braack has been working with a reference group to better understand this situation and uses the following description of such areas:

> "Transboundary conservation (and development) areas are: "Areas of land and/or sea that straddle one or more borders between states, sub-national units such as provinces and regions, autonomous areas and/or areas beyond the limit of national sovereignty or jurisdiction, whose constituent parts form a matrix that contributes to the protection and maintenance of biological diversity, and of natural and associated cultural resources, as well as the promotion of social and economic development, and which are managed cooperatively through legal or other effective means."

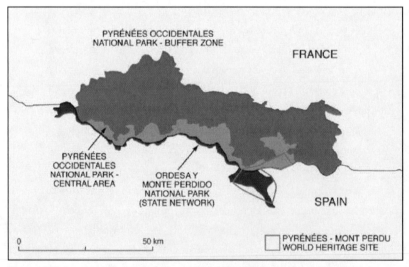

Figure 2 Pyrenees National Park (France) and Mon Perdu National Park (Spain)

We have already mentioned that some areas have been established principally to symbolize or recognize peace and friendly relationships amongst states, such as the Waterton-Glacier International Peace Park (Figure 1). The term Peace Park, though, is applied quite generally to many commemorative situations that have nothing to do with either transboundary situations or conservation at all. For the purposes of this discussion, the IUCN has defined parks for peace as:

> "Transboundary protected areas that are formally dedicated to the protection and maintenance of biological diversity, and of natural and associated cultural resources, and to the promotion of peace and cooperation."

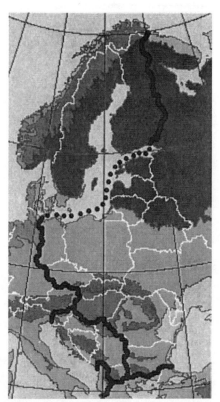

A final category includes all of those areas that are continental or even intercontinental in scale, where component parts of the landscape are linked by a biological process. For example, there is an ambitious attempt to link all of the protected areas along the former iron curtain into a continuous green line through Europe and into the Balkans (Figure 3).

Another version of this situation is when several countries agree to maintain a corridor for migratory species, such as the Agreement on the Conservation of African-Eurasian Migratory Waterbirds. The transboundary migratory corridor network includes 117 countries connecting

Figure 3 European Green Line (shown here as thick, dark-gray line)

Canada, Europe, Asia, the Middle East, and Africa. The agreement coordinates conservation action of the range states for the protection of the migratory pathway of waterbirds, in particular the protection of wetlands utilized by the waterbirds during migration. An impressive 235 species have been identified that depend on the availability of these wetlands during their annual migration cycle.

"Transboundary migratory corridors are areas of land and/or sea in two or more countries, which are not necessarily contiguous, but are required to sustain a biological migratory pathway, and where cooperative management has been secured through legal or other effective means." This situation shows clearly that it takes a wide range of sectors and interests to reach agreement to establish and manage these cooperative arrangements.

Challenges in Transboundary Conservation Initiatives

It became apparent that there was no global center or networking among transboundary practitioners and sites. A recommendation arising from the work conducted prior to the World Parks Congress was that the IUCN should take the lead in establishing a Global Network on Transboundary Conservation. This would include all organization, initiatives, and individuals involved in this topic. The idea is to link site-based practice with professionals throughout the world and develop the conceptual and applied knowledge of transboundary conservation.

This idea would involve developing a transboundary information hub, supported an Internet-based portal of contacts and resources available to all. This has now been established and is called www.tbpa.net. It also involves communicating key issues arising out of transboundary practice among individuals and initiatives. A portfolio of case studies and papers supports this goal, but falls short of the overall purpose to establish a learning network for transboundary conservation. This network would in turn necessitate the development of theory and analytical tools to support improved practice and understanding. It is hoped that through a managed program of meetings, exchanges, and interventions that the process of developing and disseminating knowledge will be advanced.

Our understanding of the global growth in transboundary conservation is that there is an enormous investment by agencies, governments, and international funders in transboundary conservation initiatives. This is not supported, however, by even minimal investment in the development of best practice guidance or in reflection on the appropriateness of varied approaches. In particular, there is a need to discriminate the purpose of the initiatives and ensure that the management approaches are appropriate. For example, if the goal is primarily biological, then the approach to implementation will support the achievement of the biological goal. But this should not be at the expense of other goals. Re-establishing a migratory elephant population can have devastating impacts on community livelihoods and security for example. If the goal is to engender peaceful cooperation, then one must be sure that the initiative by a strong neighbor does not come across as imperialistic. In essence, there is a need to refine the approaches to transboundary conservation from their current naïve status.

Finally, it is necessary that great attention be paid to governance arrangements at all levels (between the nations concerned, at the international level, at the management agency level, and at the level of local stakeholders). There is a need to analyze and refine approaches to governance, and to ensure that the emergence of transboundary conservation contributes to the development of legitimate and sustainable practice.

Tanzania Wilderness Areas

M. G. G. Mtahiko

Chief Park Warden, Ruaha National Park, Tanzania

Conservation in many of the African countries differs in terms of approaches, but it is generally accepted that no matter what system is adopted all aim at protecting the resources in an optimum condition, as would be practicably possible through application of the most contemporary acceptable methodologies. All aim to balance development that assures acceptable levels of resource impacts while taking into consideration benefit to local communities.

Proactive communities and the private sector are key dimensions to ensure this achievement in the real sense. Presently, conservation aims at enhancing satisfaction of tourists through increasingly diversified activities at a high quality with very minimum negative impact to the resources. The largest challenge, however, is to balance resources utilization with development of different facilities in line with community needs.

Conservation in Tanzania, Tanapa History

Protected Areas (PAs) were first established during the colonial era. Following independence in 1961, more conservation areas were established in several categories. Tanzania National Parks (TANAPA) are under trusteeship management, 4% of total land area of the country. Human habitation, save for park and tourism investment staff, agricultural activities, and hunting, is not allowed. Game reserves managed by the game department, 10% of total land area, are where tourist hunting is allowed. Forest reserves, under the forestry department, cover 15% of total land area, with some 3% overlap with PAs devoted to wildlife are for conservation of forests, including catchment forests. Conservation areas are where human habitation and wildlife coexists, specifically the Ngorongoro Conservation Area Authority.

Resource conservation in national parks, game reserves, conservation areas, and catchment forests is based on general management plans that are developed from Management Zone Plans. Management Zone Plans are developed essentially to govern the types and limits of tourism infrastructure and visitor use in defined zones of a PA. This paper shall concentrate on national parks, the highest level of conservation of natural resources in the country, where consumptive use of resources is not allowed.

Currently the trusteeship manages core protected areas that cover 4% of the total land area of the country. Tanzania National Parks is a parastatal trusteeship under the Ministry of Natural Resources and Tourism, enacted by an act of parliament under law of the land (Chapter 412 of 1959).

TANAPA is currently managing twelve national parks, which form the major samples of different biomes and ecological systems in the country. The organization has, through years of experience, developed a strategic planning process that is used to prepare general management and zone plans for national parks to ensure an appropriate balance between preservation and use of resources. It is mandated to "manage and regulate the use of areas designated as national parks by such means and measures to preserve the country's heritage, encompassing natural and cultural resources, both tangible and intangible resource values, including the fauna and flora, wildlife habitat, natural processes, wilderness quality, and scenery therein. The park resources should provide for human benefit and enjoyment of the same in such manner and by such means as will leave them unimpaired for future generations."

The primary objectives or purpose of national parks is to preserve areas possessing exceptional values that illustrate the natural or cultural resources of the country; areas that offer superlative opportunities for public benefit, enjoyment, or scientific studies; areas with outstanding examples of a particular type of resource; water and soil resources critical to maintain ecological integrity and that support the subsistence needs of people outside park boundaries; and to ensure:

- Parks retain a high degree of integrity as true, accurate, and unspoiled examples of a resource

- Management plans for parks are developed by interdisciplinary teams comprised of appropriate professionals with the best available information to achieve a balance between preservation and use that does not adversely impact park resources and values

- A quality visitor experience rather than "mass tourism" at the expense of park values and resources

- Optimum levels of revenue and benefits accrue to the national economy, the parks and communities, without impairing park resources

Ruaha National Park

This area was first recognized as part of the Saba River Game Reserve in 1910, which was regazetted as the Rungwa Game Reserve in 1946. In 1964, the southern portion of this reserve was declared the Ruaha National Park and in 1974 a smaller section to the southeast of the Great Ruaha River was added to complete the boundaries that exist today. Development of infrastructure has been largely restricted to the eastern central portion of the park in the Rift Valley bordering the Great Ruaha River.

A major event affecting the park's status was the completion of the bridge across the Great Ruaha at Ibuguziwa in November 1991, allowing a reliable year-round access by road from Iringa municipality some 130 kilometers (eighty-one miles) east from the park boundary. The approach to management conforms to the current policy for the preservation and management of wilderness in Tanzania's national parks. Visitor surveys in 1993/1994 indicated that the park's wilderness character was far and away the most appreciated of its qualities, and the vast majority of visitors pleaded against development that would destroy this.

The balance between use and revenue from tourism is guided by the General Management Plan (GMP), which spells out what activities and development structures can exist at a certain level in a given area. The plan thus recognizes eight zones for the purpose of resources sustainable management:

- Wilderness Zone [WZ]

- Semi-Wilderness Zone [SWZ]

- Conservation General Use North Zone [CGUNZ]

- Conservation General Use South Zone [CGUSZ]

- Core Preservation Zone [CPZ]

- Conservation Limited Use Zone [CLUZ]

- Transit Road Zone [TRZ]

- Park Administration Zone [PAZ]

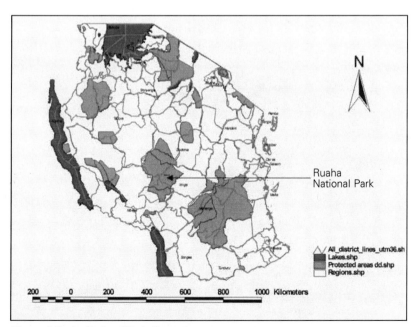

Figure 1 Ruaha National Park, Tanzania

The Wilderness Zones

"Wilderness" is a term used in conservation aspects in Tanzania. It is generally used in all forms of protected areas. Wilderness in the Tanzania conservation context refers to management of resources in a portion of a PA. Such portions referred to as wilderness zones are designated for particular uses in addition to resources conservation.

"Wilderness zone" is not a category of resource conservation in itself, but an area for specific management objectives aiming at separating various uses, mainly resources management with low impact human activities. The management of these zones are directed by a management zone plan, which spells out what activities are allowed and to what level development may or may not be done while emphasizing sustainable resource use and optimum protection.

The Ruaha National Park GMP, like in other parks and protected areas, stresses the natural quality, remoteness, and exceptional resource values for multi-dimensional visitor experiences. To achieve visitor satisfaction, it was important to establish an area that would offer visitor experience and satisfaction through walking and camping. Wilderness zones give opportunities for diversified visitor activities and at the same time offer visitor distribution, hence alleviating the problem of over con-

Figure 2 Looking up the escarpment into the wilderness area, Ruaha National Park. Photo by Michael Sweatman

centration in certain areas, but at the same time gives a chance to be close to the nature.

Wilderness Attributes

The established wilderness zone within the park has an area of 4,200 square kilometers (1,622 square miles), about 41% of the total park area of 10,300 square kilometers (3,977 square miles). This is a quite sizeable piece of land with usage set at low impact for high satisfaction to visitors. Physical development in the zone is at a minimum to offer low impact but high visitor experience and satisfaction activities. The wilderness zone provides for diversity of visitor recreation opportunities without compromising resources protection. The zone also retains qualities of remoteness while allowing a range of activities by keeping within Limits of Acceptable Use (LAU) stipulated. The area also serves as a resource bank for the future, an important component in biodiversity sustainable management. This approach of zoning clearly conforms to the current policy for the preservation and management of wilderness in Tanzania's national parks (TANAPA, 1994a).

Other Similar Areas in TANAPA and Other Protected Areas Systems

Each park that has a comprehensive GMP has an area designated as a wilderness zone where development is at a very minimum. Kilimanjaro and Ruaha's GMPs are under review, and Mikumi is in preparation. Currently the national parks, not including the newly gazetted Kitulo and Saadani, have a total area of 14,573 square kilometers (5,627 square miles) designated as wilderness zones (Table 1). Similarly, other categories of protected areas have similar setting of zones designated as wilderness.

Challenges to Management of Wilderness Areas/Zones

Despite each wilderness zone having specific management objectives, they are managed jointly as one unit with little available resources. Difficult access, their size, limited financial resources, change of polices as influenced by politics, and inadequate environmental education are the major challenges managers face.

Table 1 Park Area in Wilderness Zone

Park	Area in Km^2	Remarks
Arusha	Unknown	Zone available, but no size given yet
Gombe	33.99	
Katavi	1,000.00	1005.00 in semi-wilderness zone
Mahale	1,225.88	
Serengeti	5,149.55	
Lake Manyara	370.00	Includes the Marang Forest
Rubondo	Unknown	Zone available, but no size given
Saadani*	No wilderness zone currently	74.00 km2 semi-wilderness zone only; Newly gazetted park, GMP preparation underway
Tarangire	1,266.53	373.36 km^2 in semi-wilderness
Kilimanjaro	Unknown	GMP under review after annexure of catchment forest
Ruaha	4,200	GMP under review
Mikumi	Unknown	GMP in preparation
Udzungwa	1,327.00	
Kitulo*	Unknown	Newly gazetted park; GMP preparation procedures underway *

*Newly gazetted national parks   © TANAPA 2005

Being zones of less visitation and low development, these areas are usually at risk of illegal activities unlike those that are frequently visited. Frequent hot fires during the dry season also keep resources at risk. Illegal taking of resources, besides reducing their numbers, also disturbs the animals. This may not be a good situation for the tourists, as frequently disturbed animals tend to be aggressive.

Inadequate funding is an issue since these areas are not established for revenue generation, but management requires funds. The funds available will depend on revenue collected by the organization through services provided to visitors.

Importance of Wilderness Zones in Tourism

Wilderness zones are an important setting in protected areas as they form a core base for biodiversity conservation. Wilderness zones are also resource banks, allowing for additional diversified activities, contributing to distribution of tourists, alleviating congestion, increasing stays, and providing opportunities for conventional activities, for example, walking and hiking.

Wilderness zones provide for a diversity of visitor recreation opportunities without compromising resources protection. They retain the quality of remoteness while allowing a range of activities by keeping within the Limits of Acceptable Use (LAU) stipulated.

Park Operations

Park operations are done through linking seven departmental arrangements. Emphasis is focused on three departments: law enforcement, some aspects of outreach, and tourism.

Law Enforcement

Law enforcement is charged with protection of resources and all matters of intelligence gathering, including patrolling in the park, as well as the adjacent areas outside the park. It also performs prosecution in a court of law. Rangers in the department are in different locations that are strategically placed for optimum deployment and policing. Currently, there are seven ranger posts: Mpululu, Magangwe, Jongomero, Madogoro, Lunda, Mafinga, and Isunkavyola. The last two are still temporary, as permanent structures are yet to be constructed. Posts total eight including the headquarters. The GMP has identified one additional post in the north at Mkwambi, also yet to be in place.

These ranger posts range from fifty to more than 100 kilometers (thirty-one to more than sixty-two miles) from the park headquarters. In ideal conditions, every ranger post should have a vehicle, high frequency transceiver for long-range communication, very high frequency radio—base/car set, handheld set for use during patrols, and global positioning (GPS) equipment for ease of movement in the bush.

However, such items are not adequate as per requirements due to inadequate funding.

Tourism

The first commercial interest towards tourism development in the park was the construction of the Ruaha River Camp (now lodge), with 100 beds by Foxtreks, Ltd., at Mwayangi in 1981, some ten kilometers (six miles) upstream of the bridge. Three more tented camps are operating now, each with twenty-four beds: Mwagusi Safari Camp, Jongomero Tented Camp, and Upper Mdonya River Camp. The tented camps close during the rain season, between February/March and end of May, as there are not many visitors then.

Two facilities exist outside the park boundary; these are Tandala tented camp, which is only six kilometers (four miles) off the boundary in the east, and Tungamalenga camp in the village, thirty-eight kilometers (twenty-four miles) from the park headquarters. There is a new upcoming investment close to the village.

The park owns and runs Bandas with thirty beds, Rest House with eight beds, and Hostel with thirty-five beds specifically for schools and organized/educational groups. Booking for these facilities is necessary, as

Figure 3 A TANAPA game ranger on guard duty in Ruaha National Park.
Photo by Michael Sweatman

Table 2 Tourist Visitation in Ruaha National Park, 1996–2004

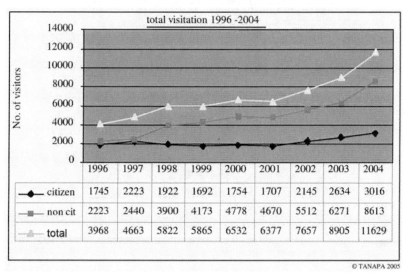

	1996	1997	1998	1999	2000	2001	2002	2003	2004
citizen	1745	2223	1922	1692	1754	1707	2145	2634	3016
non cit	2223	2440	3900	4173	4778	4670	5512	6271	8613
total	3968	4663	5822	5865	6532	6377	7657	8905	11629

© TANAPA 2005

vacancy is limited. Additionally, for visitors interested in camping, there are two public and three special campsites.

The park can be visited year round (Table 2), depending on interest and activities. However, the best times for game and sightseeing and walking safaris, both short and long, are between June and December, before the onset of rains. It is pleasing to the bird watchers to visit the park at the end of December/January through mid May, when the migrant Abdims' and yellow-billed storks can be seen. Tourism has been increasing in terms of visitation yearly. Ruaha National Park is reached by air, scheduled flights available; or by public service road, about seven to eight hours from Dar es Salaam on the coast; and from Arusha via locally registered tour operators.

Local Community Involvement and Benefit Sharing

The gazettement process of a national park starts with the local communities in the adjacent areas of the intended protected area. The communities are invited to give their opinion from the villages. The same system is adopted until the process is at the district and regional (equivalent to the province in other countries). During these stages, all

Figure 4 Feedback session with local teachers and Friends of Ruaha Society Staff. Photo by Michael Sweatman

matters that are forwarded by the communities are discussed and sorted out jointly between the regional and local government and communities.

Having been agreed to by all concerned parties, the matter is forwarded to the responsible ministry with the relevant proposals, after which the same is verified with the local authorities. With the satisfaction of the ministry responsible, the document is prepared for the cabinet to discuss all matters including the legal issues, especially on the proposed boundaries, before the bill is tabled for the parliament.

In such an arrangement, there are normally no resentments, although on some occasions the process takes a long time to ensure that all things are done correctly in the first place, understanding the communities as among the key stakeholders.

Community-Based Organization

Villages adjacent to Ruaha National Park have registered an association for the management and sustainable use of natural resources in their area. This organization (MBOMIPA—sustainable use of natural resources in Idodi and Pawaga divisions of the Iringa district) works under village governments and is supported jointly by the wildlife division, TANAPA, and the district council. This set up is an easy way for communities to take responsibility towards resources management, but at the same moment it serves as a fair way of realizing benefits accrued there from. There were nineteen villages as founding members, and requests are being received now from other villages to be enrolled as members of this association after realization of the benefits that founding members enjoy.

Issues and Challenges, Focus on Ruaha National Park

The main management problems and concerns, which the General Management Plan has sought to address, are:

- Biodiversity—There is a scarcity of dry season surface water sources for all rivers, sand rivers do flow on the surface during the rain season (mid December through mid May) but cease flowing on the surface during the dry season. Controlled use of surface water will maintain the flow, which is important for existing biodiversity.

- Endangered species—The park is endowed with different species of flora and fauna, some of which are classified by The World Conservation Union (IUCN) as endangered, such as the African hunting dog, endemic, threatened such as the cheetah, leopard, elephant, and rare. These require sound management initiatives for their survival. The core preservation zone is set to secure sensitive and fragile parts of the park along the great Ruaha River.

- Wildlife behavior—The park aims to ensure naturalness of the park through proper use of designated facilities so as to protect the animals from continuous disturbance in their habitats.

- Vegetation and soils—The park aims to control usage of surface water to sustain vegetation and maintain natural processes.

- Water resources—Continuous surface and subsurface water recharge flows are critically important in ecological processes that require constant availability. The park shall endeavor to control the usage to sustain other natural upkeep of the environment.

- Visitor experience/limits of acceptable use—These are set to ensure minimal impact of human activities to the park resources for optimal visitor experiences.

- Cultural and scenic resources—The resources will have adequate protection for continued usage by the neighboring communities and tourists.

- Neighboring communities—The park has negligible/low impact on quantity and quality of the water that runs through it. It is the obligation of the park to ensure that this is continued for downstream users.

- Park operations—The park strives to demarcate clearly all the park boundary lines for ease of recognition by the communities and other stakeholders. Research conducted shall be geared to solving resource management issues. A comprehensive resources survey will be conducted to chart where these are placed in the park.

- Revenue and tourism—The park seeks to develop game viewing facilities for economical game drives, optimum enjoyment and benefit without impairing resources, and proper administration of revenue collection.

- The Great Ruaha and Mzombe Rivers—These river systems partly form the boundary of the park. The Great Ruaha River forms the main water source for animals during the dry season (July through December). The Great Ruaha River ceases to flow during the dry season due to various uncontrolled human activities upstream of the park boundary. The park envisages working closely with other stakeholders in efforts aimed at sustaining continuous flows of the river throughout the year.

- Unique interface on miombo and east African *Acacia/ Commiphora* communities and riverine communities—This is a unique interface of vegetation communities in the park. It is aimed at protecting the species therein and prevention of introduction of species that are not common to the ecosystem.

- Significant wildlife resources—Elephants, sable, and roan antelopes, greater and lesser kudu are important wildlife species. Their abundance and unique coexistence in Ruaha is one of the park's major attractions. The park shall ensure protection of all wildlife in and around the park.

Funding

The park receives only about 30% of its base budget in the form of revenue presented to the head office. In turn, the park develops its budget like any other park. The budget for every park is jointly discussed in line with expected revenue. All the parks are regarded equally as they are all dealing with conservation, the organization's goal. The revenue collected is shared with all the parks and the head office for base costs and some funds are set for development programs and government tax. The park does get assistance in various ways from different institutions including: Friends of Ruaha Society, Friends of Serengeti, The WILD Foundation, World Wide Fund for Nature (WWF) Tanzania program office, and the Wildlife Conservation Society.

Conclusion

The park has the task of protecting resources while developing for tourism and ensuring that adjacent communities benefit from the revenues collected. There is always an issue on how to balance development for tourism with conservation. Amidst globalization, it is perhaps inconceivable to maintain areas that do not generate enough funds to meet base budget requirements. However, the organization's main goal of sustainable conservation of resources and habitats remains. All parks are of equal status in terms of conservation and needs, hence are rated on the same level no matter the amount of revenue collected.

Being in the southern part of the country, Ruaha is not well visited as compared to other parks in the north. On the other hand, this benefits the park by being visited by tourists who are more interested in nature and who do not prefer seeing many tourist vehicles. It is a place for those who need enough time to get close, watch, and appreciate nature.

Protected Areas Conservation Program at Kamchatka Peninsula, Russian Far East

Elena A. Armand

Director, United Nations Development Program, Russian Country Office

Kamchatka Peninsula is located in the Russian Far East, and still possesses globally important biodiversity that is measured not so much by species richness as it is by the presence of numerous rare and unique species, species assemblages and ecosystem processes including volcanic and geothermal ones. Approximately 15,000 Kamchatkan brown bear (*Ursus arctos*), the second largest sub-species in the world, and some 4,500 individuals live throughout the peninsula; the rare Steller's sea eagle (*Haliaeetus pelagicus*), largest in the world, inhabit there as well. Approximately 1,800 endangered northern sea lions (*Eumetopias jubatis*) are found along its coast, as does the only population of sea otters in the western Pacific. The peninsula possesses some of the world's greatest diversity of salmon, trout, and char. All species of Pacific salmon, representing one third of the entire Pacific population, spawn in Kamchatkan rivers. Nevertheless, fifty-nine faunal species on the peninsula are threatened or endangered, and are listed in the Russian Federation's Red Book.

The diversity and uniqueness previously described was the major reason why the Global Environment Facility (GEF) and the United Nations Development Program (UNDP) decided in 2001 to design and implement a conservation program in Kamchatka. The financial support was obtained from the GEF and Canadian International Development Agency (CIDA), as well as a parallel contribution from Kamchatka administration. Hundreds of stakeholders are involved in different project activities, among them are: protected areas staff, local communities, indigenous peoples associations, NGOs, research institutions, tourist operators, and associated experts from many countries (e.g., United States, Canada, Great Britain, Germany). The project is very interested in the participation of partner agencies and cooperates actively with the IUCN-World Conservation Union, the World Wildlife Fund, The WILD Foundation, and others working in the same region in order to rationalize expenses and make each input the most efficient.

The Kamchatka network of protected areas consists of territories with different protection regimes: two strict nature reserves at the federal level, seventeen purpose reserves or refuges of either federal or provincial significance, four province nature parks, one nature park at the local level, eight-three nature monuments, and other sites designated for their unique features. These protected areas (PAs), selected on the basis of various ecological characteristics, biodiversity values, and their uniqueness, comprise 27.4% of Kamchatka territory. Among them, four protected areas have been selected as project model sites. The principle of the selection was to

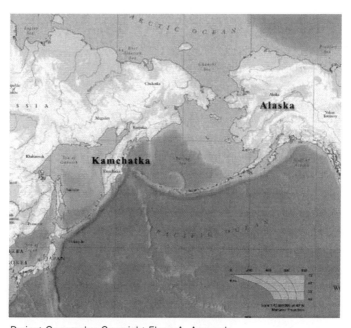

Project Geography. Copyright Elena A. Armand

combine in one project completely different territories and try to combine their advantages in management, practices, and policies. On the other hand, more advanced and well-developed areas could share their experience with "younger brothers." Before the project started, these territories did not feel like units of one conservation network and did not interact at all.

The main project goal is as follows: "To secure global biodiversity benefits by demonstrating replicable, sustainable protected areas conservation." Despite the goal itself sounding quite wide and ambitious, the five subordinated objectives are more specific and point out the way to address the goal:

1. To strengthen protected area management capacity

2. To upgrade biodiversity information and its management

3. To strengthen protected area financing

4. To strengthen legal, regulatory, and policy base

5. To heighten biodiversity awareness and advocacy

Each of these five objectives obviously create a precondition for the sixth one:

6. To develop enabling mechanisms to support alternative livelihoods and community-based conservation

The sixth objective is the main concern and focus of the whole project, as there is clear understanding among all stakeholders that global biodiversity could be secured exclusively with the participation and care of those who live within and in the vicinity of protected areas. All project components are logically subdivided into two large parts: PA-oriented activities and community-oriented activities, which could not be successful without each other. Community-oriented activities work toward the development of sustainable livelihoods, and all of these measures are directed to one end: to develop an alternative for those who are living their lives from poaching. They often do not even consider it as poaching, but quite the contrary as normal family maintenance. From the juridical point of view it is illegal fishing and hunting that should be shifted to other conservationally sound livelihoods.

In the course of the project launch and implementation, some issues and constraints were met, as in any other project. Lack of cooperation between agencies meant in some cases animosity and competition for budget or donor funds. Agencies involved could not understand why they should cooperate because they belong to a different hierarchy. In the beginning before the project, most of the protected areas staff had never been trained in management or in environmental education, and this extremely low capacity caused many challenges in fulfillment of PAs' major functions. Constant underfinancing led to speedy staff turnover, primitive lack of necessary equipment, and low public interest. Poor organization had negligible authority in society; it's a common rule. As a consequence of all previous issues, project teams discovered a very low understanding of conservation ideas among the general public, in the society living in Kamchatka, despite part of Kamchatka's nomination as the world heritage site in 1997. Moreover, indigenous communities felt offended with PA establishment, and claimed that their rights were violated.

Some achievements and outputs demonstrate, however, that efforts were not in vain. Four selected protected areas now have five-year management and annual operation plans; they had never in their history had workable planning tools. Biodiversity status is assessed and databases and maps compiled so that the PAs know specifically what they need to protect. These four selected protected areas, as well as others in the peninsula, could never before have afforded a new snowmobile or uniforms for inspectors, not to mention meet

South Kamchatcka State Sanctuary.
Copyright Elana A. Armand

the needs of the visitor center or field cabins. Now equipment is purchased and infrastructure constructed, making it possible to fulfill their direct duties. Those in the PAs only intuitively knew what to do, and had never been trained in guarding, environment education, or how to work with visitors. Training courses for PA staff and management were arranged by the project. A wide public awareness campaign included, inter alia, a so called "Environment Initiative," a social treaty signed by individuals, private companies, schools, administration, and institutions. People and organizations committed themselves to make any possible steps to conserve nature, to keep it untouched, or to restore where possible. "Make whatever you can," is the main slogan of the campaign. And as a result, thousands of people are now familiar with the necessity to conserve wilderness in Kamchatka. Hundreds of people have signed the treaty and dozens have done something already in this regard. Yet another achievement is school curricula on environment, which was developed and tested in Kamchatka so that new generations will grow as staunch supporters of the conservation idea.

To keep all these results sustainable for decades, new financial mechanisms should be invented and introduced. One such financial tool proposed is the establishment of a trust fund, or its analogue, specifically for Kamchatka protected areas. This fund is developed and registered and expected to start operations in 2006. What is already in place and functions perfectly is a small business support fund for local and indigenous communities. People living in remote rural and undeveloped settlements have, since 2003, had access to cheap and relatively easy microcredits for launching their own small enterprise. Constant consultations, trainings, simple application procedures, and assistance in reporting make it even more attractive for people. There is evidence that many former poachers prefer to establish their sustainable subsistence using this legal business model.

There are several lessons learned in the course of this project implementation, which could be of interest for those who intend to implement other projects in Kamchatka in particularly. First, full transparency is a prerequisite as a lack of understanding and knowledge of financial structures results in speculations and rumors amongst the population. Second, the expectations and concerns of local people are mainly related to the resolution of socio-economic problems while biodiversity conservation problems remain in the background. And when you appeal to local people you shall always consider their perceptions and try to look at the issue from their eyes. Third, success of any project in the field depends on community ownership and therefore a participatory approach is key to establishing this ownership. And fourth, involvement of women is crucial for success, as the livelihood patterns indicate an enormous potential for their contribution and women in local settlements are usually more active, more inquisitive in nature, and can have real influence on the mood of the whole community.

Here are additional conclusions after three years of project implementation:

1. New financial mechanisms should not be connected with budget funds. In unstable economics, no one can rely on the wisdom of the government, which will allocate increasing resources for nature conservation. This is the only way for project outputs to be kept after completion.

2. Implication of the government should be "warmed up" by clear and understandable benefits like ecosystem services, green investments, revenues from ecotourism and visitors as alternatives to very costly conditions of mining/extraction of minerals, which seems to bring quick money.

3. Federal protected areas should support local PAs in order to make a workable conservation skeleton. A protected areas network is not a commonly accepted idea in Russia, especially if we deal with federal reserves (older brothers) and local parks (younger brothers). Like in any normal family they should be united by one idea and feel that they serve the same goal rather than compete with each other.

A Perspective on Wilderness in Europe

Franco Zunino

Founder, Associazione Italiana per la Wilderness

Promoting wilderness conservation in Italy, a country with a long history of civilization and settlement, is a significant challenge. Few very large areas remain, almost none of them in a pristine state. Moreover, there is no obvious Italian equivalent to the word "wilderness," and no deeply ingrained wilderness culture as there is in countries such as the United States or Canada. Nonetheless, finding a way forward for a wilderness conservation strategy is a high priority. Italy, in particular the Alps, northern parts of the country, and in the central Appennini Mountains, can provide critical habitat for large mammals such as bears and wolves. Italy is also an integral part of the Mediterranean hotspot, and has many endemics—and unfortunately many species on IUCN's red list: twelve of thirty-nine threatened European mammal species; fifteen of the twenty-nine threatened bird species; and four of the fourteen threatened reptile species.[1] As a whole, despite its relatively small size, Italy contains more than one third of all European fauna.[2]

Despite the obvious challenges, Italy has many rural areas throughout the country that still contain wildlands, and many local populations have a deep appreciation for these areas. As a result, there is in fact a strong basis for wildlands conservation in Italy, and it has been possible to implement a gradual but highly effective strategy to start securing some of Italy's remaining wild areas, and, just as importantly, to develop a wilderness conservation ethic.

Many of the wilderness areas that have been established in Italy are small by international conservation standards, and some of these units might not qualify as wilderness in other countries. However, despite their small size, many of these areas can be expanded over time, or are already part of a larger protected complex. As such, they are in many respects building blocks: providing a foundation for larger wilderness areas to be assembled in the years to come, or, just as importantly, serving as a tool for developing a wilderness conservation culture in Italy.

For a highly populated country that does not have a culture of wilderness conservation, an incremental approach to wilderness protection is a necessity. The wilderness ethic must be nurtured, and a wilderness network must be established gradually as awareness, understanding, and acceptance of the concept grows. The good news is that this incremental approach is producing results: new wilderness areas are being established on a regular basis in Italy, providing a model that can be followed not only in new areas throughout the country, but throughout the European Union as well.

The Pizzo Madama Marta peak in the Monte Maggiore Wilderness Area.
Photo by Anna Filomena De Simone

Background

Origins of the Wilderness Movement in Italy

Italy's wilderness movement began with a booklet written by the author (then one of the staff at Abruzzo National Park, as an expert naturalist) and published by the former National Department of Agriculture and Forests (which today is divided between the Department of Agriculture and the Department of the Environment) in 1980 entitled, *Wilderness, a new necessity for the preservation of natural areas*.[3] The booklet briefly illustrated the American history of the wilderness philosophy and concept, the importance of wilderness areas, and the history of the U.S. Wilderness Act of 1964. The booklet also illustrated the first proposal for wilderness areas in Italy, as well as criteria for future European wilderness areas.

In 1981 the author began publishing and distributing a newsletter entitled *Documenti Wilderness* (*Wilderness Papers*), designed to raise awareness among Italian environmentalists of both the wilderness philosophy and the broad parameters of wilderness conservation and management. At the 3rd World Wilderness Congress in Scotland 1983, the author presented "A Wilderness Concept for Europe," during the Congress' plenary sessions, which was later included in the Congress' proceedings.[4] Momentum from the 3rd World Wilderness Congress inspired the author and several friends and colleagues to found an Italian wilderness society: Associazione Italiana per la Wilderness (AIW), which then began working to establish wilderness areas in Italy.

A creek and ravines in the Burrone di Lodisio Wilderness Area. Photo by Riccardo Tucci

Wilderness Areas in Italy

By December of 2006, there were forty-two wilderness areas covering more than 29,000 hectares (71,600 acres) in seven regions of Italy and fifteen provinces—from the Alps to the coast, to the central-southern Appennini Mountains. The very first Italian wilderness area was the Fosso del Capanno wilderness area, established in 1988, now covering 760 hectares (1,877 acres). This

A boundary cartel of the Val di Vesta Wilderness Area. Photo by ERSAF

area was first established via a management agreement over with a private foundation covering 118 hectares (283 acres). The area was expanded when the Regional Forest Authority classified an additional 259 hectares (622 acres), and then expanded again when the Municipality of Bagno di Romana added another 383 hectares (919 acres).[5] The largest wilderness area is the Ausoni Wilderness Area, which is 4,230 hectares (10,338 acres). The smallest is Brizzulera, at .3 hectares (.741 acres). Most of these areas are protected by municipalities, regional forestry authorities, or private landowners, including in some cases AIW. Only one designation is by a national park authority (Vesuvio). The largest *de facto* roadless, wild area in Italy is the Val Grande National Park at 14,700 hectares (36,300 acres). AIW played a key role in the protection of the first 11,700 hectares (29,000 acres) of this park, an effort that was strongly supported by several World Wilderness Congress resolutions.

Definition of Wilderness and Allowed Uses

AIW defines a wilderness area as an area with no roads or other industrial infrastructure, no houses or permanent buildings, no ski resorts, no wind power mills, no industrial artifacts, and no motorized use of the land. AIW therefore adopts strict protection measures to preserve the territorial integrity of the areas. However, AIW is generally open to a sustainable use of renewable natural resources, such as hunting,

The Val di Vesta drainage, with a reservoir in the foreground, in the Val di Vesta Wilderness Area.
Photo by Riccardo Tucci

fishing, gathering forest products, some logging, and grazing. With respect to logging, AIW generally does not allow any cutting in wilderness areas managed by the regional forest authorities, on lands for which AIW holds an easement or for lands that AIW acquires directly. For other wilderness areas, only very small parcels of coppice woods are clear cut, and mature forests are always logged very selectively. Grazing also generally has low impacts and in some cases is useful from a biodiversity perspective.

These criteria take into account the fact that in Italy, local people are often favorable to the idea of preserving their wildlands if protection does not mean a strict no-use policy of renewable natural resources, as it does in national or regional parks and nature reserves. This approach of respecting traditional resource use is consistent with the approach taken by many countries around the world, from Finland to Mexico, to achieve a balance between wilderness values and local uses. However, AIW always requests of the authorities who designate wilderness areas that at least part of the area must be preserved as a core area (46%) without any logging or other loss of habitat, and that at least some portion of the wilderness area (37%) must also be closed to hunting.

Designation Process

Italy's wilderness areas have not been created by legislation, but rather by internal administrative initiatives of the authorities that manage municipal, regional, or federal lands. As a result, most designations are therefore made by decrees, drafted in partnership with AIW based on the criteria above, and issued by municipalities or regional forestry authorities. In some instances, wilderness areas designated by municipal councils are then added into town planning guidelines and regulations.

There are of course some exceptions to this rule: some wilderness areas are established entirely privately by easements held by AIW or by private philanthropies, or in a few cases through direct land acquisition by AIW of wooded areas. As mentioned above, one wilderness area was established in a national park.

Direct land acquisitions, and wilderness areas created by easement, are indefinite in duration. Designations by municipal councils or regional authorities are ideally indefinite, though their status could in principle be revoked. In practice, however, this almost never happens. Almost all the wilderness areas have been approved unanimously, with support from both the majority and minority parties on the municipal councils. To date, only one municipality has ever attempted to revoke a wilderness designation (to building a wind farm) though AIW's successfully intervened to prevent this from happening.

In a very positive development, a regional Wilderness Act was has been proposed in the Lazio Region, which has the highest number of wilderness areas. Because of a change in government, the Lazio Region has not yet acted on this proposal, though AIW has hopes that this could be the first step towards a regional legislation, and ultimately

A wild aspect of the Corni di Nibbio peaks in the Val Grande National Park, saved by an AIW and WWC battle. Photo by Riccardo Tucci

perhaps a national law. Another possibility for legislation is emerging in the Friuli Venezia Giulia Region, where in December 2006 a decree[6] was passed to authorize a program for the designation of regional wilderness areas, and which identified nine areas for a total of 4,103 hectares (10,134 acres). These areas may be the most similar to the U.S. wilderness areas, they would be the first designations made by a regional government (rather than a regional land management authority), and therefore this initiative represent the highest legislative point reached in Italy to date.

Conclusion

Thirty years ago, there was almost no dialogue in Europe about wilderness areas. Certainly there was no discussion of any sort about wilderness in Italy, and very few people knew the term even existed. Today, every environmentalist in Italy is familiar with the term, a literature on the wilderness concept is developing, and experiential wilderness trail programs are gaining in popularity. In 2005, the Italian government officially recognized the AIW as an official environmental preservation association through a decree from the Department of the Environment. And some organizations are even beginning to speak about the necessity of a wilderness areas concept by national law.

There is much work yet to do, both at the policy level and in terms of designating new wilderness areas in Italy. Nonetheless, it is important to take stock of the successes to date, and to the fact that we have successfully adopted the philosophy of Aldo Leopold, who referred to a wilderness area as: "A continuous stretch of country preserved in its natural state, open to lawful hunting and fishing, devoid of roads, artificial trails, cottages, or other works of man."[7]

CHAPTER 10

The United States

"Leaving Alaska"

~

CAROLYN KREMERS

Honorable Mention,
8th World Wilderness Congress Poetry Contest

I begin to understand
the reasons
I cannot seem to shake
this place
from the eye of my desire

it is a streaking sky

it is a thin birch

it is a bird song at night

and it is a fish, swimming *upriver,*
reflecting light
safe return
the change of seasons
shiver of wind

and the white and black
moth that lands
like a Yup'ik word
on the triple-paned glass
of the cabin window
and tilts its patterned wings,
intricate and slim

marked
in the slanting sun
it flickers again

swimming
flicks again

Untrammeled Wilderness

Douglas W. Scott
Policy Director, Campaign for American's Wilderness

"The wilderness that has come to us from the eternity of the past
we have the boldness to project into the eternity of the future."
—Howard Zahniser,
Architect of the Wilderness Act

A Growing System of Protected Areas:
Statistical Overview

The National Wilderness Preservation System in the United States comprises 106,619,199 acres (43,147,259 hectares) as of January 15, 2006, in 680 units in forty-four of the fifty states.

The Wilderness Act of 1964 protects these diverse wildlands to secure for "present and future generations the benefits of an enduring resource of wilderness." Moreover, they are protected by the strongest legal mechanism available in the American system of government—by statutory law. These preserved wildlands amount to 4.7% of all the land in all ownerships in the United States.

This record of accomplishment contrasts sharply with the record of federal wilderness preservation before the Wilderness Act became law. Most of the agencies that administer portions of America's federal lands had taken little if any initiative to preserve wilderness areas. The U.S. Forest Service pioneered in establishing wilderness areas by agency administrative order in the 1920s and 1930s. However, over the twenty-five years prior to the signing of the Wilderness Act on September 3, 1964, the total acreage protected by this increased by just 2%, to a modest 14,617,461 acres (5,915,477 hectares; see Table 1). And these original areas were protected only by agency administrative orders, which could easily be altered by the agency, rather than by statutory designation.

With enactment of the Wilderness Act, Congress immediately gave the much stronger protection of statutory law to a portion of those earlier administratively protected areas, totaling 9,139,721 acres (3,698,714 hectares). Since then, Congress has passed more than 125

Table 1 Historic Progress of Wilderness Protection in the United States			
Agency	**As of January 1940**	**As of September 1964**	**As of January 15, 2006**
U.S. Forest Service:			
Wilderness areas (administratively designated before 9/64; statutory after)	0	9,139,721	34,877,591
Primitive areas for study (given interim statutory protection by Wilderness Act)	14,235,414	5,477,740	173,762
U.S. Forest Service—Total	14,235,414	14,617,461	35,051,353
National Park Service	0	0	43,650,796
U.S. Fish and Wildlife Service	0	0	20,699,108
Bureau of Land Management	0	0	7,391,704
Total Acres Statutorily Protected (wilderness *and* primitive areas)	**0**	**14,617,461**	**106,792,961**
Number of States with Areas	13	13	44
Boundary Changes Can be Made	By administrative decision	Only by act of Congress	Only by act of Congress

Notes and Data Sources

Primitive area acreage from U.S. Forest Service. These received statutory interim protection in the Wilderness Act, so are counted here as statutorily protected only after September 1964. Only the Arizona portion of the Blue Range Primitive Area remains under this interim statutory protection. January 2006 data from www.wilderness.net (except Blue Range primitive area, from website of Apache-Sitgreaves National Forest.) Millions of acres of lands in all four agency jurisdictions are pending in presidential recommendations not yet acted upon by Congress, or are in official wilderness study areas, and these categories are not counted in this table.

additional wilderness designation laws, adding 97,479,478 acres (39,448,545 hectares) to the national wilderness system, nearly twelve times as much!

And the wilderness system of the United States continues to grow. The most recent addition was a 100,000-acre (40,469 hectare) wilderness area added by a law signed by President George W. Bush on January 6, 2006.

Thus, forty-one years after the enactment of the Wilderness Act, that historic law can be declared to be an outstanding success and an extraordinary testament to the wisdom of those who, building on the vision of Leopold and Marshall, conceived the ideas of such a law and worked for its enactment by the U.S. Congress.

Why the Wilderness Act?

The United States has the good fortune to possess a very large expanse of public lands remaining in the custody of the federal government, one-third of the entire country. Despite the inroads of advancing human use, a significant portion of these public lands remain substantially unroaded and wild. The pioneering thinkers of the American wilderness movement sought to protect some of these wildlands as wilderness areas and to find a means of doing so that would most securely protect them as wilderness in perpetuity.

This line of thinking began with Aldo Leopold, who envisioned a system of wilderness areas encompassing several categories of federal lands, including the wild portions of national parks and national forests, among others. He proposed that such areas be established by agency administrative order. His efforts led to the first such designation, the Gila Wilderness in 1924. In the 1930s, Bob Marshall took up Leopold's idea, advocating for protection of more lands and stronger protective policies. Working as officials within federal land management agencies, it was natural that both men initially urged protection by means of agency administrative orders. Changing priorities and politics within these agencies could, and soon did, lead to trimming back of earlier wilderness designations to make way for development.

Given their goal of preserving wilderness areas in perpetuity, it became increasing evident to wilderness protection advocates that reliance on agency administrative orders would not and could not assure that the protected areas would remain so.

By the mid 1930s, Leopold, Marshall, and other leaders came to doubt whether federal land management agencies could be relied upon to honor these preservation decisions over a sustained time period. One leading conservationist published the stark conclusion that, regarding both wilderness within national parks and national forests:

> "There is no assurance that any one of them, or all of them, might not be abolished as they were created—by administrative decree. They exist by sufferance and administrative policy, not by law.

Through the 1940s, the idea for a wilderness preservation law, an act of Congress, evolved from this growing disenchantment with reliance on protection promised by agency administrative orders. In the American governmental system, a law enacted by Congress offered a stronger guarantee, both that the protective policies would be sustained and that the boundaries of individual wilderness areas, once established, would not be likely to be reduced.

(NOTE: The parenthetical references to "book" provide page numbers for relevant sections of the author's book, *The Enduring Wilderness: Protecting Our Natural Heritage through the Wilderness Act*, Doug Scott (Fulcrum Publishing, 2004, www.fulcrum-books.com) that expand on each topic and provide further references.)

The Key: Statutory Protection

Acts of Congress are very difficult to enact. The Founding Fathers deliberately designed the legislative process with an inherent inertia, requiring success at each of many procedural steps and thus giving an advantage to opponents of proposed legislation. For just this reason the Wilderness Act, first introduced in 1956, took eight years to become law. The architect of the legislation, Howard Zahniser, observed that an:

" … Outstanding characteristic that I have learned to emphasize in our congressional government—in our whole government—that is this: It is very difficult for anybody in our form of government to get anything done that anybody doesn't want done. Now you can see right away, that's a pretty good characteristic of a large democratic government established by a people who have learned to fear tyranny and to fear over-government."

Under the Wilderness Act, a new act of Congress is required to add any land to extend its protection to additional land. This places the burden on those who advocate wilderness protection for additional areas. Yet, that very burden is the single most important fact about American wilderness preservation policy, for the Wilderness Act also specifies that the boundary of a wilderness area can only be changed by another act of Congress. Thus, once an area has been protected by Congress, the burden of the inherent legislative inertia shifts to those who might seek to weaken its protection or alter its boundary. Such changes have occurred about a dozen times over forty-one years, but all have been minor and in most cases involved correction of errors in the original boundary (book: 15–18).

The Politics of Wilderness Preservation

The authors of the Wilderness Act did not come from the federal agencies or Congress, but from the citizen conservation movement. Indeed, both Leopold and Marshall were founders of The Wilderness Society, for they understood the need for strong advocacy from outside the government. In 1947, this organization resolved to initiate a campaign for a nationwide system of wilderness areas to be protected by a wilderness law. They also understood that the nature of congressional politics carried with it inherent implications:

- Legislation will best progress if it has broad and bipartisan support within Congress.
- Congress is moved by many forces, but strong grassroots citizen advocacy from across the country is essential.

- Congress does not enact sweepingly visionary legislation, but acts incrementally, through "the art of the possible."

- Where legislation appears to impact one state or locality, other members of Congress give by far the overwhelming say about that legislation to the members of that state's congressional delegation.

- The essential lubricant that allows any law to be enacted by a diverse legislative body is accommodation and compromise.

These, then, are the immutable pragmatic realities of legislative politics (book: 114–122). They have come into play with every one of the more than 125 laws Congress has enacted over four decades to add lands to the National Wilderness Preservation System. And they came into play in the design and details of the original Wilderness Act itself.

To some wilderness enthusiasts, the very idea of compromise is anathema. They have a very pure vision of wilderness and, in many cases, strong distaste for the realities of legislative politics. Fortunately, those who conceived and drafted the Wilderness Act, led by Howard Zahniser executive director of The Wilderness Society, took a more practical approach. In this they continued a fundamental pragmatism that was always central in the thinking of Leopold, Marshall, and the other pioneers of the wilderness preservation movement.

Nonconforming Uses in Wilderness Areas

It was perfectly obvious to Leopold and his colleagues that they were seeking to apply wilderness protection to lands on which other, sometimes conflicting, uses had already been long established and practiced. As he wrote in 1925:

> "An incredible number of complications and obstacles … arise from the fact that the wilderness idea was born after, rather than before, the normal course of commercial development had begun. The existence of these complications is nobody's fault. But it will be everybody's fault if they do not serve as a warning

against delaying immediate inauguration of a comprehensive system of wilderness areas."

Similarly, Bob Marshall understood that "certain infringements on the concept of an unsullied wilderness will be unavoidable in almost all instances," observing that, " … almost all the disadvantages of the wilderness can be minimized by forethought and some compromise." As he drafted the Wilderness Act, Zahniser followed this line of thinking, as he wrote:

> "Where considerations of expediency or recognition of existing practices have permitted inconsistent wilderness use—such as domestic stock grazing within designated wilderness areas in the national forest system—such uses should be recognized as non-conforming and looked upon as subject to termination as soon as this can be done and done equitably for those immediately concerned. Such nonconforming uses should be permitted only when their temporary sufferance appears to be a means of insuring future values of the area."

The assertion is sometimes made that if the law designating a wilderness area includes special management provisions allowing nonconforming uses, congress is allowing activities that degrade that wilderness area. However, when Congress designates a new wilderness area, the act of Congress it passes does not cause those nonconforming uses. Rather, they already exist as physical realities on the ground—and as political realities, too (book: 130–131).

As he introduced the first version of the Wilderness Act in 1956, Senator Hubert H. Humphrey (D-MN) explained the philosophy that guided the approach to this problem in the legislation:

> "Existing uses and privileges are respected in this bill, and private rights are protected. This is not essentially a reform measure but rather a measure to insure the preservation of a status quo which fortunately includes a great resource of wilderness.

"Special provision is made for the protection of existing rights and privileges on any areas involved. Grazing within the national forest areas is provided for as at present, and existing uses authorized or provided for in (national wildlife) refuges are also permitted. The termination of nonconforming uses is provided for whenever this is agreeable to those making the uses."

"Purity" and the Wilderness Act

How "pure" must an area be to be considered for designation as a wilderness area? That is, beyond the presence of some nonconforming uses, must the ecosystem itself have been untouched by man? Here, too, the designers of the Wilderness Act followed the guidance of Aldo Leopold, who wrote in *A Sand County Almanac*, "Many of the diverse wildernesses out of which we have hammered America are already gone; hence in any practical program the unit areas to be preserved must vary greatly in size and in degree of wildness."

And how did the Wilderness Act deal with this question? Many miss the fact that subsection 2(c), the definition of wilderness, comprises two sentences: the first an ideal characterization, the second a deliberately more practical definition. In 1961, Senator Clinton P. Anderson (D-NM) was both the lead sponsor of the Wilderness Bill and chairman of the Senate committee handling the legislation. In opening hearings that year, he definitively explained his legislative intent in these two definitions of wilderness. The first sentence is a definition of pure wilderness areas, where "the earth and its community of life are untrammeled by man." It states the ideal. The second sentence defines the meaning or nature of an area of wilderness as used in the proposed act: A substantial area retaining its primeval character, without permanent improvements, which is to be protected and managed so man's works are "substantially unnoticeable." The second of these definitions of the term, giving the meaning used in the act, is somewhat less "severe" or "pure" than the first.

Commenting on the two-part structure of the definition during the final Senate hearing on the Wilderness Act in 1963, Zahniser noted that:

"In this definition the first sentence is definitive of the meaning
of the concept of wilderness, its essence, its essential nature—a
definition that makes plain the character of lands with which the
bill deals: the ideal. The second sentence is descriptive of the areas
to which this definition applies—a listing of the specifications of
wilderness areas; it sets forth the distinguishing features of areas
that have the character of wilderness."

The first sentence defines the character of wilderness; the second
describes the characteristics of an area of wilderness.

The practical effect of this distinction, as repeatedly underscored
by Congress as it shaped and explained subsequent wilderness designation laws, is that "impure" lands showing the impact of human uses,
such as past human settlement, are not barred from protection under the
Wilderness Act (book: 66–72; 126–129).

Stewardship for a Perpetual Wilderness System: The "Non-Degradation Principle"

The goal of projecting wilderness into the eternity of the future requires
more than a Congressional decision to designate areas; it takes great care
by the agencies administering these special places. As one close confi-
dant of Zahniser's as he drafted the Wilderness Act wrote:

"At the same time that wilderness boundaries are being established and protected by acts of Congress, attention must be given
to the quality of wilderness within these boundaries, or we may
be preserving empty shells."

The practical challenges of preserving wilderness character are
today the work of thousands of devoted employees of the four federal
land management agencies—administrators, planners, experts in diverse
resource fields, and on-the-ground wilderness rangers. The Wilderness
Act sets out one overarching directive for these wilderness stewards
regardless of which agency is involved:

"Except as otherwise provided in this act, each agency administering any area designated as wilderness shall be responsible for
preserving the wilderness character of the area and shall so
administer such area for such other purposes for which it may
have been established as also to preserve its wilderness character."

Zahniser linked this fundamental command to the ideal concept
in the first sentence of the act's definition, which would otherwise have
no function in the act, an assumption not allowed by the rules of statutory interpretation. As Zahniser told Congress, the ideal definition
makes plain the character of lands with which the bill deals and thus
functions to give meaning to the act's command that administrators preserve "wilderness character."

The allowance or prohibition of various uses within wilderness
areas is central to the protections of the Wilderness Act. A subsection
of the law captioned "Prohibition of Certain Uses" carefully distin-
guishes uses that are flatly prohibited—commercial enterprises and
permanent roads—from others that may be allowed for agency personnel under very limited circumstances, including the use of
chainsaws, motorized equipment and vehicles, among other things.
The agency discretion to use these tools is limited in one of the most
delicate of Zahniser's phrasings:

"Except as necessary to meet minimum requirements for the
administration of the area for the purpose of this act (including
measures required in emergencies involving the health and safety
of persons within the area), there shall be no temporary road, no
use of motor vehicles, motorized equipment or motorboats, no
landing of aircraft, no other form of mechanical transport, and
no structure or installation within any such area."

Senator Frank Church (D-ID), a leader in passing the Wilderness
Act, explained the rationale for this provision:

"We intend to permit the managing agencies a reasonable and necessary latitude in such activities within wilderness where the purpose is to protect the wilderness, its resources, and the public visitors within the area—all of which are consistent with 'the purpose of the act.' The issue is not whether necessary management facilities and activities are prohibited; they are not. The test is whether they are in fact necessary."

The elegance of this so-called "minimum requirement" language lies in its two-part test for allowing exceptions for use of normally prohibited tools by agency personnel themselves. First, is the proposed exceptional activity necessary and second, if so is the means proposed the minimum tool to achieve that purpose? The federal training center for wilderness stewards, the Arthur Carhart National Wilderness Training Center, provides a detailed "Minimum Requirement Decision Guide" to equip wilderness stewards with instructions and worksheets to apply this minimum requirement analysis to proposed actions, projects, and activities in wilderness areas.

Taking the command to preserve wilderness character and the two-part wilderness definition together, the result is a non-degradation directive to wilderness stewards. Congress may, and does, designate lands that are less than pure, lands with some fading imprints of man's work or existing, nonconforming uses, for these are within the meaning of what Senator Clinton Anderson called the "somewhat less 'severe' or 'pure'" second definition in the act.

The point of the non-degradation principle is that whatever an area's past history of human impact, once a wilderness area has been designated the goal for stewardship is to manage that area toward the ideal concept of the first definition, so that "the earth and its community of life are untrammeled by man," so that the wilderness is to the greatest extent possible what the etymology of the word suggests: self-willed land.

Wilderness in a Larger Context

Deepening ecological sophistication brings even greater challenges in stewardship of wilderness. For example, the issue of invasive non-native species poses conundrums regarding how far, if at all, wilderness stewards should go in intervening to manipulate the environment toward the goal of restoring what we perceive to be more natural conditions. Yet, in a world warming under the impact of global climate change, that are other man-caused ecological change occurs so pervasively that sorting man-caused change from natural ecological progression may even prove impossible.

The very fact that our ecological sophistication is growing could tempt us, at any given moment, to conclude that we have enough wisdom to know what well-intentioned manipulation is best to achieve some preconceived wilderness ideal. However, in my view this temptation to ecological hubris conflicts at the most fundamental level with the self-restraint and ecological humility bound up in the very idea of wilderness.

Howard Zahniser believed that well intentioned manipulation of wilderness could pose a serious threat to the wilderness concept itself, leading to "rationalization of projects to carry out certain current concepts of … management" conflicting with "the wilderness philosophy of protecting areas at their boundaries and trying to let natural forces operate within the wilderness untrammeled by man." As the ideal, he said, wilderness areas "should be managed so as to be left unmanaged, without man's management or manipulation." "With regard to areas of wilderness," he wrote, "we should be guardians not gardeners."

This paper is a summary of the material presented in Mr. Scott's recent book, *The Enduring Wilderness: Protecting Our Natural Heritage Through The Wilderness Act.* Fulcrum Publishing, 2004.

Wilderness and
the National Park Service

Fran P. Mainella
Director, National Park Service

In the United States, the National Park Service shares responsibility for the National Wilderness Preservation System with the Bureau of Land Management, the U.S. Fish and Wildlife Service, and the National Forest Service. While our agencies have different missions and purposes, we have a shared commitment to wilderness protection.

Under U.S. law, our country has formally designated 105 million acres as wilderness, of which the National Park Service manages 41%, or about 46 million acres. The majority of those acres are within the national parks in Alaska.

Too often, in America, it is presumed that "wilderness" is about exclusion, locking lands up and keeping people out. It is not, and will not be!

Of course, we see wilderness in terms of protection, but we also see it in the context of people and their enjoyment. In Alaska, much of our wilderness is managed to accommodate the needs of people whose legacy includes living off the land. In the contiguous forty-eight states, we are more restrictive in our management. The difference is not because wilderness there is more fragile—Alaska's tundra, for instance, is very fragile—and certainly not because that wilderness is "better," since the inspirational power of Alaska's wilderness landscapes is clearly the equal of any.

But, we recognize that wilderness is conceived by people and managed for people. We preserve wilderness not for its sake, but for our and our posterity's sake. We need what wilderness gives us. That necessarily means its management reflects the needs of people and must be adjusted to recognize place, condition, and use.

The concept of wilderness and the role of humans in the wilderness landscape continue to evolve. Twenty-five years ago, the U.S. Congress passed the Alaska National Interest Lands Conservation Act (ANILCA), creating vast areas of wilderness here in Alaska. Recognizing the critical importance of these lands to the lifestyle of Native Alaskans and other rural inhabitants, congress made provisions in ANILCA to more clearly establish the appropriate continued presence and subsistence traditions of humans in Alaskan wilderness. It was a wise choice.

We must value and protect wilderness. We must honor it. But we cannot deprive it of the dynamic relationship with people. This administration came to office with several commitments regarding our national parklands:

- The first and most often discussed is the commitment to address the massive maintenance backlog that we inherited.

- The second was our determination that, through our Natural Resources Challenge, science would be fully integrated into our management of parklands.

- To ensure that people would continue to be able to enjoy their national parks through appropriate use that will leave the park resources unimpaired for present and future generations.

The Wilderness Act of 1964 was developed and debated for eight years. When it came to the floor of Congress for a vote it passed by the amazing vote of 374 to 1, demonstrating the very broad, unified support of the American public. It was also the first federal law to mandate public involvement in the designation and management of protected lands, pre-dating the National Environmental Policy Act by five years.

When that historic legislation passed, Congress declared that wilderness areas should be devoted to several public purposes: recreation, education, conservation, as well as scenic, scientific, and historical use.

The job isn't done. The National Park Service has more than 27 million acres of land that are still being considered for possible designation as "wilderness." Congress, the courts, and the public expect that we will continue to be good stewards of those lands, protecting their wilderness character until a determination is made. Collaborative planning with ample public involvement will enable us to meet this challenge.

We have our own challenge for Congress, too: Our Wilderness Action Plan asks Congress to seriously consider the nineteen park wilderness proposals that remain before it.

After lands are designated wilderness, we have the responsibility for protecting them in perpetuity, while providing for their use and enjoyment by the public. To help us address these responsibilities, I have included the implementation of the Wilderness Action Plan in our National Park Service Legacy Goals.

Within the park service, we have a National Wilderness Steering Committee representing all of the professional disciplines involved with wilderness stewardship from all across the country. The committee works diligently to address and resolve the challenging issues of protecting the wilderness ecosystem, while providing outstanding opportunities for the public to enjoy their natural and cultural wildland heritage.

Because wilderness management in the U.S. is a multi-agency task, we also participate in an Interagency Wilderness Policy Council, currently chaired by Karen Taylor-Goodrich, Associate Director for Visitor and Resource Protection with NPS. It is a measure of our commitment that we have one of our top-level Washington managers directly involved in the effort to assure the U.S. has a truly unified National Wilderness Preservation System. The council and a staff-level interagency steering committee collaborate to assure consistency, continuity, and accountability for wilderness stewardship among the partner agencies.

The National Park Service continues to train park personnel and develop handbooks to guide wilderness stewardship planning and management in parks. We have a continuing emphasis on scientific monitoring and research to better understand both the natural dynamics of wilderness and the potential for improving of the human lives it inspires by experience.

Wilderness ecosystems and the human experiences within them are complex and diverse. The wilderness landscape provided the setting and forces that shaped and molded the American character, and helped defined who we are today. The deliberate act of setting aside examples of such wildlands, as a legacy to future generations to use and enjoy, was an idea first developed by Americans.

The greatest challenge of any land management is the management of human activity. The restlessness of human imagination guarantees such management is always tested. Our job is to accept that challenge and assure that our management is thoughtful.

Saying "no" is easy, but unproductive. Saying "yes" requires vigilance, but opens the rewards of the wilderness to all people, not just the chosen few. If humans are treated as unwelcome non-native species in our wilderness, they will soon become as unconcerned as they are unwelcome. We in the National Park Service believe the right choice is not always the easy one. Ultimately, the success of any park program depends on cooperation with an informed public. To be informed, we must provide public access to the treasures of wilderness.

We have the opportunities to impart the value of the wild and the legacy the land. But the skill of the woodsman is taught. The dynamics of the desert must be learned. We cannot depend on people's instincts to give them an appreciation of the heritage we protect. And we cannot expect them to love what they cannot see or experience.

If the next generation is to share our appreciation for wilderness values, they must first be experience them. And, they must be taught those values. We can do that. Let's work together to make it happen.

The Gwich'in People, Oil, and the Arctic National Wildlife Refuge

Luci Beach
Member, Gwich'in Steering Committee

I want to thank the aboriginal people of this area especially from Eklutna, for allowing us onto their land. I would like to say three things for you to think about: Home, heart, and sacred. I very pointedly was told recently that the word "wilderness" is inappropriate at times, because what we can refer to as wilderness is actually home. "Home" is land to people and has been for a millennium. And that is what the North Slope of Alaska is to our brothers and sisters of the North, the Inupiat. "Sacred" is what the Arctic Wildlife Refuge calving and nursery grounds are to the Gwich'in, and that has been the case for millennium. "Heart" is what the Gwich'in have with the Porcupine caribou herd and what the Porcupine herd has with the Gwich'in.

Caribou are the lifeline of the Gwinch'in culture. Seasonal caribou hunting camp with outlook tower. Photo by Gwich'in Steering Committee

In our creation story, it is told that the caribou would always retain a part of the Gwich'in heart, and the Gwich'in would always retain a part of the caribou heart. *Iizhik Gwats'an Gwandaii Goodlit*, this is what we refer to the calving and nursing ground: The Sacred Place Where Life Begins. At one time, the Gwich'in were nomadic, and then after contact our villages were located strategically along the route of the caribou herd. In my own family I have stories of three famines of which the Gwich'in survived, and not once did we go to the calving grounds because it is our belief that in order for us to continue, that the caribou at that time have to be left alone to regenerate.

What we're talking about is drilling in the heart of the national wildlife refuge. Yes, maybe a small part, but it's the heart where birds from all fifty states and six continents go for nesting and staging. And to us the most critical part is where 40,000 to 50,000 caribou calves are born. Recently, because of global climate change, the herd had severe difficulties making it to the calving grounds. We lost thousands of caribou calves because their instincts are so strong to cross the Porcupine River, which they are named after, and because the weather had changed—the ice wasn't there when they needed it, so many calves died.

I would like to make the distinction, too, between being a member of the Gwich'in Nation and being a shareholder of a Native corporation. My grandchildren are not members of the corporation that I belong to; they are members of the tribe I belong to. Sixty to 70% of

Gwich'in women performing traditional dance. Photo by Gwich'in Steering Committee

our diet comes from the land. I also need to tell my brothers and sisters, and other tribes that are here form the lower 48 states, that the corporations are not the economics of the tribe. They are separate entities. They do not answer to the tribe.

When our elders and our leaders first heard about the possibilities of development in the refuge, there was a gathering called at Arctic Village. I should tell you anytime there's a threat to the Gwich'in, from time immemorial, we gathered as a people. So people from all the corners of the Gwich'in Nation gathered in Arctic Village, and it was at that time we determined to oppose development of the Arctic National Wildlife Refuge. The Gwich'in Steering Committee was formed to go forward and educate people about the Gwich'in and that this is a human rights issue that we are allowed to continue our way of life that we've known since time immemorial.

We've been villainized. We've been characterized. We have been used by this faction and that faction. But the elders of two rules told us to do it in a good way, not to scream or point at people who oppose us, and to speak from the heart. They wisely told us and we've tried the best we can, with very little money and not a lot of people. But the people on the Steering Committee over the years have done an incredible job, and we've aligned our self together with people in the environmental community. So we are considering them our friends. We get most of our support from other tribal people (as well as people in the faith community), because other tribes, particularly in the lower forty-eight, remember what they had at one time: the great herds of buffalo that got in the way of manifest destiny when we had the policy in this nation of genocide, which later on, after we went under the Department of Interior Policy, became one of assimilation.

For us this is a serious situation, and I have to say to that we support our brothers and sisters to the North for no off-shore development because we understand that's what their culture is based on. This is a serious time. Rather then putting the Arctic Refuge into a straight up or down vote, proponents have taken the backdoor approach. This is not honest. We must maintain integrity and protect the Artic National Wildlife Refuge.

Wilderness Insights from Alaska:
Past, Present, and Future

Deborah L. Williams

President, Alaska Conservation Solutions

Alaska contains abundant wilderness areas and intact ecosystems that have been thoughtfully protected as a result of remarkable efforts from literally thousands of people. It is a pleasure to share the history of these wildlands accomplishments, in the hope that our experiences may benefit conservationists, agency personnel, and indigenous peoples throughout the world. As in other states, provinces, and countries, however, Alaskans are also facing significant threats and challenges to our natural heritage. We know that many of these threats, such as global warming, require coordinated national and international action. To protect Alaska and other great natural places in the future, we must work together and learn from one another.

Background

For more than 100 years, men and women with extraordinary vision have legally protected a significant percentage of Alaska's magnificent natural areas. Currently Alaska, which is greater than 365 million acres in size, contains more than 40% of its land in varying degrees of protected status. Most significantly, Alaska includes more than 57.9 million acres of Congressionally designated wilderness, which represents more than 50% of all designated wilderness in the United States. This wilderness is contained in many of Alaska national parks (33.49 million acres), national wildlife refuges (18.67 million acres), and in the Tongass National Forest (5.75 million acres) (BLM, 2003).

More generally, Alaska's national parks protect 54 million acres (more than 65% of all national parkland in the U.S.); Alaska's national wildlife refuges protect 70.7 million acres (more than 80% of all national wildlife refuge land in the U.S); and Alaska's national forests contain 22 million acres (representing the two largest national forests in the U.S.) (BLM, 2003). There are other nationally protected areas in Alaska, including 3,131 miles of wild and scenic rivers and two Bureau of Land Management (BLM) protected areas. In addition to federally designated areas, the State of Alaska has placed some of its most ecologically significant land into state parks and state refuges, totaling 11.3 million acres, with 6.5 million of those acres designated as state parks, refuges, sanctuaries, and critical habitat areas (Hull and Leask, 2000).

Alaska's conservation success story, however, consists of much more than the number of acres that have been protected. Notably, we have sought to honor human relationships with Alaska's lands and waters. This is especially reflected in the subsistence laws and practices that are authorized on federally protected lands (ANILCA Title VIII) and in the broad recreational and other use of these lands. The laws governing Alaska also recognize, both explicitly and implicitly, the important economic benefits, ecosystem service benefits, and existence values of protected lands for present and future generations (ANILCA Title I).

As a result, Alaska's intact ecosystems are providing tremendous cultural, economic, and social values. Alaska's rural indigenous peoples are able to subsistence hunt and fish on tens of millions of acres of lands and waters, sustaining one of the most important foundations of their heritage. At the same time, more than 26% of all jobs in Alaska rely on healthy ecosystems representing more than 84,000 jobs and providing $2.5 billion a year in wages and even more contributions to the economy overall (Colt, 2001). Key economic sectors that require healthy ecosystems include commercial fishing and processing, tourism, subsistence, recreation, and government employment. In Alaska, our wild places also contribute tremendously to quality of life and our identity, as reflected in polls showing that more than 90% of Alaskans believe a healthy economy is necessary for a strong economy and more than 70% identify themselves as conservationists (Moore, 2001).

Achieving the Protection of Alaska's Wildlands

There have been four primary factors in the successful protection of Alaska's wildlands. While individually all of these considerations have been important to the safeguarding of Alaska's lands and waters, these factors have also complemented and reinforced each other. The four factors are: 1) thoughtful stewardship by Alaska's indigenous peoples; 2) bold protection actions by elected and appointed officials; 3) supportive science and scientific analyses; and 4) broad, positive, coordinated, generous, and strategic public and NGO engagement.

Thoughtful Stewardship by Alaska's Indigenous Peoples

To begin with, Alaska was tremendously fortunate to be well stewarded by its First Peoples for thousands of years. Living in close connection with the lands and waters, Alaska Natives harvested fish and game sustainably, while leaving few permanent traces and no apparent ecosystem disruptions. When the first non-Natives came to Alaska in the 1700s, they experienced robust populations of fish and wildlife and unscathed ecosystems. Because of Alaska's remoteness, few non-Natives occupied it until the mid 1900s, and Alaska's indigenous peoples continued to demonstrate outstanding stewardship.

Bold Protection Actions by Elected and Appointed Officials

The United States purchased Alaska from Russia in 1867. Fortunately, shortly thereafter numerous great political leaders, such as Theodore Roosevelt, took bold actions to protect areas in Alaska. From the beginning, leaders thought in terms of protecting millions of acres; they had a large vision, not small. Among other great places, President Roosevelt protected the Tongass National Forest and the national forest closest to Anchorage, the Chugach National Forest. This represents the second basis of Alaska's conservation successes: farsighted, significant, bold legislative and administrative actions by brave political leaders.

Similarly, when he fought for the passage of the Alaska National Interest Lands Conservation Act (ANILCA), President Carter was absolutely instrumental in achieving more than 100 million acres of protected lands in the state. Intrepidly using the Antiquities Act of 1906, President Carter achieved the administrative protection of more than 50 million acres of land until Congress acted, while at the same time Secretary of Interior Cecil Andrus also exercised his regulatory powers through the Federal Land Management Policy Act (FLPMA) to protect additional tens of millions of acres of Alaska lands. At the congressional level, several great senators and members of the house of representatives, including Senator Paul Tsongas, Representative Morris Udall, and Representative John Seiberling, heroically sought national engagement in the protection of the "last frontier" (Cahn, 1982). They knew that to achieve significant land protections at an ecosystem scale they needed to hear a mandate from throughout the United States, and they did.

In the end, Congress passed ANILCA and proclaimed: "It is the intent of Congress in this act to preserve unrivaled scenic and geological values associated with natural landscapes; to provide for the maintenance of sound populations of, and habitat for, wildlife species of inestimable value to the citizens of Alaska and the Nation, including those species dependent of vast relatively undeveloped areas; to preserve in their natural state extensive unaltered Arctic tundra, boreal forest, and coastal rainforest ecosystems; to protect the resources related to subsistence needs; to protect and preserve historic and archeological sites rivers, and lands, and to preserve wilderness resource values and related recreational opportunities including but not limited to hiking, canoeing, fishing, and sport hunting, within large Arctic and sub-Arctic wildlands and on freeflowing rivers; and to maintain opportunities for scientific research and undisturbed ecosystems (ANILCA, 1980, Section 101[b])." Without doubt, this is one of the boldest, most inspiring statements ever made by a legislative body regarding the protection of wildlands.

Supportive Science and Scientific Analysis

To justify, and in some cases spur many of the great conservation protection actions, Alaska has benefited from outstanding scientific studies, analyses, and advocacy by agency and academic personnel. This third factor in Alaska's success story cannot be overemphasized. The articulate voices of wildlife, wilderness, and ecosystem experts have been instrumental in achieving the conservation of such extensive areas. Armed with scientific knowledge and ecosystem health sensibilities, men and women in government agencies and universities have often lead the way, insisting on large area protections and providing crucial data and mapping capabilities.

Broad, Positive, Coordinated, Strategic, and Generous Public and NGO Engagement

Especially in the last fifty years, Alaska has also benefited from broad, coordinated, and strategic public engagement, at both the local and national levels. This has been the fourth basis for Alaska's conservation successes. As noted above, the support for Alaska's protection has been widespread throughout the United States, and has also been at both the grassroots and the "grasstops" levels. In other words, there have literally been millions of Americans who have urged Congress and the White House to protect Alaska, while at the same time there have been a large number of individuals with significant political influence who have done the same.

It is noteworthy, indeed, that so many Americans, from Florida to Washington and from Maine to California, have taken time out of their lives to write letters, call their congressmen, and even travel to Washington, D.C. on behalf of Alaska. It speaks volumes to the deep passion people have to protect the Nation's—and one of the world's—last great wild places. It also reinforces the importance of the word "national" in Alaska's national parks, national wildlife refuges, and national forests. Everyone in the country has a stake in the future of these publicly owned and managed natural places; and as a result, everyone has a voice that needs to be considered when making decisions about the protection and management of these lands.

Numerous effective conservation organizations have helped inform and coordinate these efforts, supported by the contributions of thousands of members. Repeatedly, individuals, businesses, and foundations have generously donated to the protection of Alaska. This generosity, coupled with the effectiveness of non-profit conservation organizations, has been instrumental in Alaska's conservation success story. At the same time, having a mix of national, state, and local organizations working together in a coordinated fashion has also been essential. In this regard, the Alaska Coalition has served a crucial function in coordinating and communicating with more than 400 groups in the late 1970s, and more than 900 groups today.

It is also important to note that ANILCA and many other great conservation achievements in Alaska have only been realized because conservationists have worked with Alaska Natives in recognizing subsistence rights and needs. Subsistence uses are defined as: "the customary and traditional uses by rural Alaska residents of wild, renewable resources for direct personal or family consumption as food, shelter, fuel, clothing, tools, or transportation; for the making and selling of handicraft articles out of nonedible byproducts of fish and wildlife resources taken for personal or family consumption; for barter or sharing for personal or family consumption; and for customary trade (ANILCA Sec.803)." Title VIII of ANILCA explicitly protects the rights of rural Alaskans to engage in subsistence hunting and fishing on most federal lands.

In the end, of course, conserving great natural places takes time, perseverance, focus, optimism, and dedication. Conservation representatives, scientists, Alaska Natives, politicians, and others spent more than ten years to achieve the passage of ANILCA. Other areas have taken even longer. Is it worth it? The answer to this question is definitively yes, not only for present generations but also, especially, for future generations.

Sustaining Conservation Victories and Protections

As every conservationist knows, achieving legal protections for wildlands is the beginning, not the end, of the journey. Conservationists must continue to be vigilant in monitoring against subsequent legal encroachments and detrimental management decisions. For the most part, Alaska conservationists have been successful in this regard for five major reasons: 1) constant vigilance; 2) supportive funding; 3) outstanding coordination and communication among conservation groups; 4) making the case for protection; and 5) creating champions.

Constant Vigilance

First, there has been constant vigilance. Instead of relaxing after an area has been protected, Alaska-based and national conservationists have maintained their focus. We often use the phrase, "constant vigilance, constantly applied." There has always been so much at stake, and so many powerful challengers and threats. While this has meant no rest for the weary, it has also meant that virtually all of Alaska's conservation victories have been maintained.

Supportive Funding

To sustain this level of effort has required a steady flow of funding. Fortunately, individual contributors from across the U.S. have understood this, and continued to be loyal and generous. Far-sighted foundations and businesses have also been instrumental in funding the analyses, outreach, litigation, administrative work, and communications necessary to uphold the initial victories. In this regard, it is important to remember that to obtain funding you must ask, year after year, while presenting a compelling case.

Coordinated, Collaborative, Cooperation among Conservation Groups

One way to make a compelling case is to demonstrate that groups working to protect an area are working together in a complementary fashion, without duplication of effort. In Alaska, there has been excellent coordination, communication, and collaboration among conservation groups at both the state and national levels. This is the third factor for success, and a very important one.

Making the Case

It is also crucial that advocates for protection continue to make a compelling case for safeguarding an area, and broadcast that case effectively using all available means of communication. Alaska conservation advocates have done this well, utilizing websites, list serves, film, books, member magazines, and newsletters, while also working closely with all aspects of the media. As we learn more about the importance of natural systems, and as these systems become more scarce and valuable, we have an even stronger case for their continued protection.

Creating Champions

Finally, Alaska has been successful in maintaining its protected lands because we continue to create new champions in three ways. Conservation organizations and others have conscientiously brought decision makers to Alaska to show them firsthand what is at stake. There is no substitute for a direct experience, whether it be standing on fragrant, spongy tundra while hundreds of caribou run across a nearby hill or catching a salmon on a pristine icy river while eagles fly overhead. Alaskan conservationists have also recognized the importance of cultivating youth and nurturing their leadership skills and enthusiasm, both at the high school level and college level. Through college intern programs and an outstanding project called Alaska Youth for Environmental Action, a diverse group of young people are bringing fresh enthusiasm and insights to the efforts to defend Alaska's wildlands.

Major Threats

Going forward there are many serious threats to Alaska's wildlands and intact ecosystems. Even though Alaska is experiencing some of these especially intensely, these are threats that, if unaddressed, will jeopardize wildlands throughout the world.

Global Warming

First and foremost there is global warming. No place in the United States has warmed more than Alaska: over 3 degrees Fahrenheit in the last four decades, with our winters warming over 5 degrees Fahrenheit during that same period. The adverse ecological and human impacts are being felt everywhere. These have been well documented by scientists, Alaska Native elders, and others (ACIA, 2004).

The dramatic, adverse consequences of global warming are too extensive to list in this paper, but I will highlight a few. Alaska has experienced unprecedented forest die-offs from spruce bark beetles and other insects and diseases. In the last two years, Alaska has had dramatically increased forest fires, with the 2004 season breaking all records when more than 6.5 million acres burned. Ecologically critical ice sheets are thinning and retreating; while on land, permafrost is melting. Lakes are drying up, reducing habitat for migratory birds and other animals. Many animal species are at risk, most notably polar bears and other ice-dependent species like ring seals. These and other species are at serious risk of being displaced or losing their habitat altogether. If extensive warming continues, extinctions for many species are easily predicted.

All of these global warming consequences affect not only the plants and wildlife in Alaska, but also the people who rely on those plants and animals for their subsistence way of life. Humans are also adversely impacted by global warming from the spread of diseases to crumbling infrastructure, all of which will cost millions of dollars to address. Then there is inundation.

An important lesson is simply that Alaska—right now—is unequivocally demonstrating the dramatic, adverse effects of loading our atmosphere with an excess of greenhouse gases. We must cap emissions immediately if we are to save our Arctic and other ecosystems in the long run. This is a national and international imperative.

Two Other Threats

I will briefly list two other major threats that Alaska, in particular, is facing: persistent toxic chemicals, and nineteenth-century thinking versus twenty-first century solutions. Unfortunately, persistent organic chemicals migrate from many different countries to Alaska, using a diverse array of transporters including air, water, and animals. These toxics are bioaccumulated in higher trophic species such as orcas and polar bears, raising concerns about the reproductive capabilities of these species and human consumption. Like global warming, the transport of toxics underscores our international inter-dependence on one another, regardless of where we live, and the widespread consequences to our wildlands of behaviors that are destructive and unsustainable.

Alaska also suffers from politicians and others who engage in nineteenth-century thinking instead of twenty-first century solutions. The rush to drill the coastal plain of the Arctic National Wildlife Refuge is a prime example of this thinking. Instead of spearheading efforts to conserve energy, promote energy efficiency, and enhance alternative energy production, politicians like President Bush, Vice President Cheney, and Alaska's congressional delegation are stuck in the nineteenth century's mindset of drilling for oil, which is especially unacceptable when that oil is underneath ecologically significant lands. Drilling for oil in the Arctic National Wildlife Refuge or in other protected wildlands will not have any meaningful impact on America's energy needs or on America's energy dependence, since the United States has only 3% of the world's oil reserves, yet consumes 25% of the world's oil. If we are to bequeath our children an earth rich in wildlands, we can and must demand that our elected officials engage in twenty-first-century thinking

I would also like to note an overall concern about the risk of detachment from our wildland roots. Similar to so many other places in the world, America is becoming more urbanized. If people become disassociated from (or even worse alienated from) wilderness, then it will be harder to justify the protection of wildlands based on experiential, spiritual, and existence value rationales.

Opportunities

Fortunately, there are many opportunities to affirm and expand the protection of wildlands. I will briefly discuss three: 1) the growing understanding about the values and economic importance of protecting wildlands and biodiversity, and the costs of not doing so; 2) the renewed commitment to future generations and intergenerational equity; and 3) expanded work with diverse constituencies.

Protection Values

As people throughout the world watched the devastation from Hurricane Katrina, it became increasingly clear that this catastrophe was both a natural and a human-enhanced disaster. The wreckage in Katrina's wake underscored, in dollars and immense suffering, the costs of destroying wetlands, forests, and natural rivers, together with the costs associated with global warming. In other words, the graphic lesson from Katrina and several other recent disasters is that destroying wildlands and disrupting ecosystem services is very expensive in real dollars and human misery.

As economists and others expand their understanding about monetized values of ecosystem services, it is critical that this be communicated clearly to the general public. We need to publicize success stories as well as explain catastrophes, using teachable moments wherever possible. The insurance industry can be an important ally in this effort. I believe that we are nearing a tipping point regarding public understanding about ecosystem services.

In addition to ecosystem services, it is equally important to value direct and secondary employment benefits from wildlands. There is such excellent work being done in this regard throughout the nation and the world. As noted earlier in this paper, the Institute of Social and Economic Research (ISER) examined this issue and determined that approximately 26% of all jobs in Alaska are related to intact ecosystems, generating more than 84,000 jobs and providing $2.5 billion a year in wages (ISER, 2001). Personally, I presented a speech throughout Alaska touting these statistics, and received an amazingly positive response from chambers of commerce and rotary clubs. Indeed, we must first generate this information and disseminate it widely.

Intergenerational Equity

While it is very important to describe and discuss the economic benefits of wildlands protection, it is equally important to explain our intergenerational equity responsibilities. Very simply, what kind of world are we leaving for our children and grandchildren, and how does that compare to the world that we inherited from our ancestors? As a corollary, do we have any right to give future generations a diminished planet, a planet with fewer wildlands, less biodiversity, compromised ecosystem services, and less wilderness? By any ethical standards, the answer is no. Accordingly, we must be responsible stewards, not hedonistic consumers of Earth's sustainable and life-giving bounties. For example, the spiritual, intrinsic, ecological, and other values of wilderness can only be bequeathed if these areas are carefully protected by us.

We should proudly talk about intergenerational equity. This is a fundamental value that is critical to the long-term survival of our species. We have the opportunity and the need to invoke this tremendously important value more frequently.

Diversification

To protect ecosystems and wilderness areas in the long-run, we must insure that a majority of people agree that safeguarding wildlands is appropriate, or better yet, highly desirable. As a result, we must continue to diversify the ethnic and cultural support base for ecosystem protection.

Every culture has its special relationship with the natural world. To recognize this, celebrate it, and share it with others is important. In Alaska, we created the "Guide to Alaska Cultures," as a starting point for understanding the histories and relationships different cultures have to our state and its wildlands (Alaska Conservation Foundation, 2004). It has been a very successful publication, and is now even used by the Anchorage school district as a text for the required course "Alaska Studies."

Conclusion

As wildlands become scarcer and more valuable, the job of protecting wilderness areas is more important than ever. In closing, I want to thank everyone who is dedicating their lives and careers to safeguarding wildlands. Whether it is through science, economics, advocacy, management, governance, writings, cultural understanding, or photography, every effort strengthens the likelihood that future generations will be able to experience the extrinsic and intrinsic values of wildlands that are essential to our survival, well-being, and identity. There is much to do, but together we will succeed. We must.

References

Alaska Conservation Foundation. 2004. Guide to Alaska's Cultures. Anchorage, AK: ACF Published. 90 p.

Alaska National Interest Lands Conservation Act (ANILCA). 1980. Public Law 96-487. December 2, 1980.

ACIA. 2004. Impacts of a Warming Arctic: Arctic Climate Impact Assessment. Cambridge University Press. 140 p.

Bureau of Land Management (BLM). 2003. Alaska Land Distribution Statistics (current to 2003). Division of Conveyance Management. Anchorage, AK.

Cahn Robert. 1982. The Fight to Save Wild Alaska. Audubon Publishing. 33 p.

Colt Steve. 2001. What's the Economic Importance of Alaska's Healthy Ecosystems? R.S. No 61. University of Alaska Anchorage Institute of Social and Economic Research. 4p.

Hull Teresa, Leask Linda. 2000. Dividing Alaska 1867–2000. UAA ISER Vol. XXXII, No. 1:6.

Moore Ivan. Alaska-wide Poll. Anchorage Alaska. 2001.

The Next Steps in Wilderness
Solutions and Strategies

William H. Meadows
President, The Wilderness Society

It was a lively and opinionated discussion that gave birth to The Wilderness Society. In 1934, four men, three of whom worked as government foresters, were on their way to a forestry conference near the little town of Norris in my home state of Tennessee. They became so agitated discussing the Forest Service's inability to protect wilderness areas that they pulled off by the side of the road and wrote out the charter for The Wilderness Society. We were officially founded the very next year, in 1935. This just goes to show what passion for the land, a group of good minds, and a powerful unity of purpose can lead to.

As I envision the future, I also reflect upon those early days of The Wilderness Society, three decades before passage of the seminal Wilderness Act of 1964 and when the idea of having nearly 106 million acres of designated wilderness gracing this nation would have been considered a pipe dream. We have come a staggeringly long distance from that afternoon in Tennessee. But we still have many miles to go before our journey is complete.

I'll begin with the reason of our work: wilderness. What is the value of wilderness on its own terms? Simply put, wilderness sustains our lives and spirits, teaches us to be better human beings, and is a living conservatory of our national soul. Generally when conservationists evoke the idea of permanently protecting land under the designation called "wilderness," we highlight its more utilitarian benefits: cleaning the air, purifying water, maintaining critical habitat, offering space for recreation and personal renewal. Its life-giving properties are endemic to survival on Earth.

The values missing from this equation are more subtle, but just as real. When we can temper our egotism, stand humbly before creation. and let nature take its own course, we are learning to live cooperatively with the planet and ultimately with each other. Wilderness teaches us this truth about ourselves and, like the Smithsonian Institution, safeguards a portion of our collective soul, embodying both our history and our future. Our national character was forged out of the wilderness, which represents unfettered freedom, hope in the renewing forces of nature, limitless horizons, creativity, spontaneity, and rugged individualism—a reflection of all that we call American.

Any consideration of wilderness must necessarily embrace four key ideas: people, places, policies, and politics. And when I talk about people, I am really referring to the idea of a sustaining community. As the conservation giant Aldo Leopold once noted so eloquently: "All ethics … rest upon a single premise: that the individual is a member of a community of interdependent parts. … The land ethic simply enlarges the boundaries of the community to include soils, waters, plants, and animals, or collectively, the land."

This nation, divided into groups the media calls red and blue, seems to have lost sight of Leopold's conservation ethic. The current political divide is a real philosophical breach, rooted in a fundamental difference in emphasis on the good of the commons, the community, versus an emphasis on the good of the individual. For us, wilderness is the tangible expression of community.

Wilderness connection to soils, waters, plants, and animals is obvious. Less well known but equally important is the nexus of our wilderness areas with people. In the summer of 2005, I attended the twenty-fifth anniversary celebration of ANILCA in Anchorage. One of the great triumphs of that law was its recognition of the historic and life-sustaining relationship that Native communities have with the land. Alaska has 231 distinct tribes, many of which depend on the wilderness to fish, hunt, and support their timeless ways of life. For example, the Gwich'in, whose name means "people of the caribou," have joined with us in countless skirmishes over protecting the Arctic National Wildlife Refuge because the refuge is where the herd of 150,000 caribou travel each year to calve and feed their young. Without the caribou, this community that is thousands of years old would not survive physically, culturally, or spiritually.

Let's not limit community to subsistence, however. Anyone from the State of New York is no doubt aware that New York City decided some time ago to forego spending billions of dollars to build a new water treatment plant and instead put money into preserving the watersheds that served the city. Wilderness areas, and the potential wilderness found in places like national forest roadless areas, are the nurseries for healthy watersheds and much, much more public good.

We need look no further than Hurricane Katrina to understand this question of community. Our government's emergency response system failed the people of New Orleans. And we failed that community as well by failing to be good stewards of our natural systems, the wetlands, and barrier islands that would have protected it. More often than not it is the poor and disenfranchised who suffer the most when we recklessly misuse nature. Witness the folks fishing for dinner along some of Florida's inland waterways where warning signs describe the potential for mercury poisoning. A commitment to social justice is finally a commitment to the community of life to which we all belong. And wilderness is the indispensable heart of that community.

So what are the people strategies to achieve more wilderness protection in a red-and-blue world? First, we need to broaden and diversify our activist base by creating new alliances, building new partnerships, and meeting people where they are. When I think of wilderness and people, an image that comes to mind is of Bart Koehler who runs our Durango-based Wilderness Support Center. Bart is standing on a dusty roadside in Nevada in 100-degree heat with a huge map spread out on the trunk of his car. He's talking to a couple of ranchers who want to see where the wilderness boundaries will be. After many years spent building trust and sometimes almost being run out of town, we now have two new allies nodding their heads in agreement.

Latinos from the barrios of Los Angeles, African Americans residing in downtown Atlanta, and Minneapolis soccer moms all have a stake in sustaining the community of life because that is what sustains them and their children. We must find a way to ask them to stand alongside us.

Second, we need to reclaim the moral high ground for land conservation, making wilderness protection a family value. Tragedies like Katrina may be minimized if we have the humility to live knowing that Mother Nature will care for us if we care for her. Finally, we must link people to the land by moving away from the harsh rhetoric of the past to tell a better story, a tale about humanity's inexorable connection to the natural world. And that is where all the scientists in the room come in. They are crucial to our understanding the second item on my list: place.

The Wilderness Society is an organization of place. We learned long ago that while people respond positively to the notion of wilderness; they are truly moved by those special places that awaken their senses, stir the emotions, and uplift the spirit. People connect to place. We could take a magic marker and draw some lines on a map and call it "place." Or, we can explore its depths to discover the true worth of a particular landscape to our citizens.

America's wildlands represent one of the greatest, hidden stashes of uncounted wealth in this nation and we are wasting them at an appalling rate. What is the value of a natural system that cleans our air, filters our water, or sequesters carbon? Can anyone even determine a price tag for the rare plant that might someday help cure cancer? Science is the key to unlocking nature's manifold mysteries. And we need more of it.

Isn't it galling then to have an administration playing fast and loose with the scientific process? Let me be crystal clear: I am not disparaging those dedicated and highly professional scientists who work for the federal government. They do incredible research despite difficult political and budgetary constraints. But I am calling into question officials who misuse, abuse, or edit science merely to score political points.

Our community needs more ecologists, economists, biologists, sociologists, geographers, archeologists, and geologists to build up a storehouse of intellectual capital in service to the planet. Our wilderness champions on Capitol Hill need facts to back up the cases they put before Congress. Our grassroots partners need them to convince local media, rural business owners, and elected officials. And every American citizen needs sound information to understand exactly what this nation could lose if we squander its natural treasures.

Those treasures include the countless other creatures who share this planet with us. Science continues to teach us about the priceless value of habitat. The question is whether we will take those lessons to heart and, like the biblical figure Noah, build a bigger wilderness ark for the wildlife that need safe havens against tempests and vagaries of this fast changing, twenty-first-century world?

I tip my hat to all of the environmental organizations, agencies, and academic institutions whose research is changing the face of conservation today. This intersection of science and place and activism create salvation. In fact, The Wilderness Society is so concerned about loss of potential wilderness-quality lands in Alaska, that we are establishing a new Center for Conservation Science here, whose mission is to verify the places that are important to protect and to guide our decisions on how to make conservation happen in the forty-ninth state.

My focus has been on science under the strategies for place, but let me add a few more. We need to be on our guard and defend every acre we've protected by responding powerfully whenever greed starts to encroach on our wilderness. We need to be visionary and we need to stay engaged. Many times our work really begins after the actual wilderness acreage has been designated, when a series of poor management decisions could cause a landscape to die a death by a thousand cuts. Finally, let's use science to not only make our case for protection but to help restore the land where and when we can. All of these tactics point me to the final two items on my alliterative list: policies and politics.

Someone once said that you have to be a die-hard optimist to be in this business. I am. But I must also candidly admit to having spent many a sleepless night since January 2001, worrying about what might happen next and how we could possibly bring our wilderness vision into reality.

The good news is that wilderness is still largely a bipartisan issue. In the last three congresses, fourteen bills designating new wilderness areas have been passed and signed into law, adding approximately 2.4 million acres to the National Wilderness Preservation System. The bad news is that the task of defending our gains while trying mightily to affect policy on every front, legislative, administrative, and judicial, is almost too daunting to contemplate.

Wilderness historian Doug Scott has called wilderness designation the, " … most democratic land allocation process we've invented." And therein lies our hope. Wilderness arises from persistent citizen advocates who do everything from taking inventories by literally walking the land, to responding to an electronic Wild Alert, to writing letters to the editor. I do not believe that we will have a conservation majority in 435 congressional districts any time soon, but I do believe we can develop a powerful conservation voice in those districts, the voice of passionate, committed people who care deeply and will choose to act.

The strategies for changing the politics and policies surrounding wilderness lies in our ability to: 1) reshape the conservation dialogue in this country by finding common ground with diverse communities of people and inviting them into our movement; and 2) in our ability to keep the rule of law in place and the processes democratic. An ill wind has swept through the halls of Congress and the agencies, trying to slam the door on public participation in federal land use decisions. We cannot and will not let that happen.

The United States divided into red and blue. It is a nation full of spirited people who have a dab of green in their hearts. Wilderness is not a game of them versus us. It is a mirror into eternity that reflects back the best of us; this community of life to which we all belong.

One of my favorite people was my friend and colleague, former Senator Gaylord Nelson, the founder of Earth Day. Gaylord worked as The Wilderness Society's counselor for almost twenty-five years before he passed away in July 2005 at age eighty-nine. "If we human beings learn to see the intricacies that bind one part of a natural system to another and then to us, we will no longer argue about the importance of wilderness protection, or over the question of saving endangered species, or how human communities must base their economic futures not on short-term exploitation but on long-term, sustainable development. If we learn, finally, that what we need to "manage" is not the land so much as ourselves *in* the land, we will have turned the history of American land-use on its head."

CHAPTER 11

The Wilderness Perspective—
Moving Forward

"Wild Gifts"

~

IAN MCCALLUM

"Ecological Intelligence," *Africa Geographic*, 2005

To begin
to know wilderness,
something in me had to die—
the pregnant parts,
the motherly expectations
and the test tube notions
of a safe delivery.

In the wild
dead fetuses are for real,
vultures are the midwives of new life
and to be abandoned is to grow.

To being
to know wilderness,
something in me had to come alive—
my wild side,
the part that knows
that it is impossible to sleep with the dead
without being awakened by them.

In the wild
the animal spirits are for real
they are the shadows in our bones
and they come to us
as wild gifts.

It's a Wonderful World

Bittu Sahgal
Founder, Editor, *Sanctuary*

"I see skies of blue and clouds of white
The bright blessed day, the dark sacred night
And I think to myself what a wonderful world."
—Louis Armstrong (George Weiss/Bob Thiele)

Ishmail, the game ranger, led us through his South African forest as a priest might show us his temple. "Here you can see where a hyena rested a while. Look! In the droppings is a hoof of an impala. Walk silently and the forest will speak to you."

Just before we reached the river, home of Nile crocodiles and hippos, the ever-smiling Isaac, also a game ranger, a friend, and guide, gently pointed to the ground: "*Imbewu!* The seed. See it sprout from the dung of the hippo? And this is the 'wait-a-while tree' that has caught my shirt in its thorns."

We rose early after sleeping out in the cold of the bush at the *Imbewu* camp in the Kruger National Park, a facility dedicated to providing previously disadvantaged African children with wildlife experiences. With us on a daybreak trail were four young "Kids for Tigers" ambassadors from India: Prithvi from Delhi, Shruti from Chennai, Varsha from Dehradun, and Nishant from Mumbai. All under thirteen, they were like fresh blotting papers, sponges soaking up information, experiences, and purpose. Keeping them company were some of the brightest young children on the planet who lived in Soweto, and were part of the youth program christened *Imbewu,* founded by the Wilderness Foundation (South Africa) and run in partnership with South African National Parks (SANPARKS). These children were the future of South Africa. They were the future of the world.

Ever so softly Nishant whispered to me: "When I walk in wild places I feel alive. It's exactly what I want to do all my life." All the young naturalists on whom my hopes and those of hundreds of wildlife defenders are being pinned echoed his feelings.

"Want to learn about managing waste in the city? Just look at the dung beetle. It turns shit to life," said Anish Andheria, naturalist with *Sanctuary,* a photographer, and a passionate believer in kids.

"Close your eyes. Allow the earth and its spirit to seep into you. You are safe and you belong." That was Noel de Sa, mentor and guide to us all, besides being the national coordinator for Kids for Tigers, the *Sanctuary* Tiger program that encouraged 1 million Indian children to pause a while and contemplate their place in a world still populated by tigers.

Earlier, at the botanical gardens in Pretoria, the charismatic Murphy Morobe, ex-Chairman of the SANPARK'S board, chairman of the 7th World Wilderness Congress (2001), and our host, welcomed the

Kids for Tigers in South Africa with *Imbewu* youth, Anish Andheria.
Photo courtesy of *Sanctuary*

Indian kids to Africa, saying: "We are bonded; this is the nation that shaped Mahatma Gandhi. India is very special to us and so are you young tiger ambassadors. If you work together with these bright young children of Africa, you will be able to save the wildlife of both our countries and the human cultures that have evolved from our wildernesses." He spoke with passion about the *Imbewu* program and the hopes that the elders, including Nelson Mandela, had for young South Africa.

Imbew, the seed. What a perfectly wonderful term to describe everything I have strived to achieve all my life; to seed future generations with the love and respect for the earth, vital to their survival and that of millions of species, including the tiger. While an ignorant, arrogant generation of shortsighted adults stampeded over Earth's fragile beauty, we had to somehow protect it and change the ambitions of those destined to inherit the planet. And we had to sow seeds of hope, which I did by gently reassuring the children: "You are children of Mother Nature. Like the cut on your elbow or knee, she can magically heal wounds inflicted on the earth. The turtles and crocodiles will purify your rivers. The elephants, rhinos, and leopards will help your forests to regenerate. Anemones and polyps will restore bleached corals to health. And the birds will cast fruit seeds all around to re-green your lands. But, naturally, if you keep worrying and scraping the wounds, neither your elbow, nor the earth, will be healed."

The Environmental Prophet

Mohandas Karamchand Gandhi was born 138 years ago on October 2, 1869. Educated in India and London, he pursued a career at the bar, where acute shyness almost ruined his chances of success in the earliest stages. By the time he was thirty, he was well established in South Africa, but found it difficult to stomach the way colored people were being treated by the government of the day. In protest, he gave up his law practice around 1900 to fight against the biased legislation. Within five years, he saw that the system could not be fought from within, so he opted out, gave up the Western way of life, and forsook material possessions to lead by example.

He fought valiantly for years for the well-being of his people in South Africa, using the simplest and most effective means to counter a powerful foe—*satyagraha*, or non-violent civil disobedience. He calculated, correctly, that the South African government of the day would be unable to respond to the power of peaceful resistance and got them to agree to repeal anti-Hindu discrimination.

He returned to India in 1915 and joined the freedom movement. During World War I, Gandhi the tactician supported the British in the hope that this might help convince them to free India. But this was not to be. A retinue of broken promises and massacres saw hundreds of innocents butchered and forced him to launch a series of non-violent protests against British rule.

A phenomenal motivator, Gandhi was eventually able to weld a disparate country together in joint purpose. He led India to freedom. When he died, the politicians of India government swore to uphold his ideals.

That promise was soon forgotten. It is still forgotten.

In 1947, Mahatma Gandhi told Jawaharlal Nehru that India should not chase the illusion of Western development because such dreams were built on the presumption that cheap resources to fuel material ambitions would come from other countries. He pointed out that if all Indians were to aspire to such as lifestyle, several planets would be needed to feed their demands. His kernel advice is even more relevant today in a world on a self-destruct mission:

"Stay independent. Keep your consumption and demands low. Ask first if your plans will benefit the poorest, weakest Indian before you implement them. Let the villages determine their own destiny for they are the womb of India."

Unfortunately Prime Minister Nehru, though he loved Gandhi deeply, felt this was impractical. He, therefore, created a system that encouraged educated or well-connected Indians to step neatly into the British jackboot.

The process of stripping India bare of its natural wealth, which the British had begun centuries ago, continues apace, with rich and powerful urban Indians usurping the resources of the rural poor. Today, for instance, water for 15 million citizens in India's financial capital, Mumbai, comes from distant forests and the clamor to drown still more forest to feed insatiable demands rises. Our electricity comes from dams built on the properties of villagers who were never compensated for their lands or houses. Mining and timber operations eat into their forests from the Himalaya to the Andaman. Our toxic wastes poison the aquifers that supply their wells.

Because their homes, forests, and fields were systematically stolen or degraded, millions of Indians began to stream into cities. Many still populate slums where they must take up tough, underpaid jobs. The overcrowding of urban India is a direct result of the fracturing of rural India. And the resultant pollution and environmental degradation robs both rich and poor of the quality of life guaranteed by India's constitution.

An environmental prophet, Gandhi was probably wasted on India's freedom. His teachings and his leadership could have delivered us from the environmental nemesis towards which *Homo sapiens* seems so resolutely headed. Were he alive he would surely have pointed out that even more serious than the erosion of our soils is the erosion of our value systems.

Intergenerational Colonization

I am an Indian and proud to be one because I live in a land whose ancestors respected the earth. The vast majority of Indian still venerate the earth and its myriad life forms. But we have been infiltrated. Instead of exporting our earth-loving attitudes, we continue to import false ambition's broadcast from 1818 H Street, Washington, D.C., the headquarters of the World Bank. And the agents of the destruction of our subcontinent are the very politicians in whose hands Gandhi trustingly placed the mantle of freedom. British colonial ambitions were immoral. But what the leaders of today are doing is far more immoral than that. They are colonizing the hopes, aspirations, and security of the unborn.

This is what Gandhi wrote soon after India gained her independence, as he watched in horror how a dream had gone sour:

> "I have a few letters describing some of the dishonest means congressmen are resorting to in order to further their selfish interest. I do not want to live to see all this. But if they go on deceiving us, there will be such a tremendous upheaval that the golden history of our cherished freedom, won without shedding a drop of blood, will be tarnished."

Had the lines been written yesterday they could not more accurately describe the betrayal of tomorrow at the hands of the likes of President Bush of the United States, or Dr. Manmohan Singh, Prime Minister of India. Both are in denial of climate change. Both are moving the planet closer to the precipice.

It is all too obvious that the teachings of Gandhi have been forgotten in the land of Gandhi's birth. Decades after his death, the virus

Prime Minister Dr. Manmohan Singh with children from Kids for Tigers, the *Sanctuary* Tiger Program. The banner quotes an ancient Indian saying: "The forest is the mother of the river." Photo courtesy of *Sanctuary*

of self-interest contrives to destroy India's fabled wealth that conquerors and colonial forces were unable to exhaust.

To put it simply, India has decided to sell its family jewels to some of the most predatory financial forces in the world. Thus Orissa's water-stocked forests and turtle-populated seas are hostage to iron ore companies, Gujarat's pristine coastline is being pillaged by petroleum interests, Andhra Pradesh's thick forests are being mined for uranium, Karnataka's Western Ghats are under assault by dam builders, Madhya Pradesh's tigers are being forced to retreat before invading industrialists, and fragile Himalayan glaciers, together with Earth ice everywhere, are in advanced stages of melt.

India has some of the finest environmental laws in the world. It is also a democracy. This is why the Supreme Court of India has consistently upheld environmental appeals against the destruction of our forests, often castigating the most powerful leaders in the country for their shortsighted ambition. Such politicians have not seen Al Gore's *Inconvenient Truth*, but they epitomize the despair contained in the quote of Winston Churchill that Gore used to such telling advantage:

> "The era of procrastination, of half-measures, of soothing, and baffling expedience of delays is coming to a close. In its place, we are coming to a period of consequences."

Had Indian politicians seen Gore's film, they might have realized that in an era of advanced climate change it was suicidal to castrate India's Forest (Conservation) Act, 1980 and its Wildlife (Protection) Act, 1972 by passing the new and lethal The Scheduled Tribes and Other Traditional Forest Dwellers (Recognition of Forest Rights) Act, 2006, which (ostensibly to benefit forest dwellers) is a thinly disguised ploy of politicians to counter supreme court judgments by dismantling the protective laws that prevent the powerful from trading in wilderness real estate for cash and votes.

If Tomorrow Comes

Today, in India and across the world, forests, estuaries, mangroves, wetlands, grasslands, mountains, and even deserts—ecosystems that should be jealously protected to sequester carbon in the decades ahead—are being set upon by commercial forces that have historically snatched land from the poor and unempowered, in whose name the lands have now been transferred.

Ironically, these were the very assets that Gandhi wished to save from the clutches of the British for the security of the children of the Ganges. It saddens me to see how far India has drifted from the teachings of Gandhi, who reminded us that, "A worthy heir always adds to the legacy that he receives."

Mohandas Karamchand Gandhi will have died in vain if we do not wake to the realization that the erosion of our soils is a direct result of the erosion of our value systems. "The demands of equality supersede the letter of the law," he chided the British, when they attempted to take shelter behind one-sided legislation.

Would that he were alive to repeat the advice for the benefit of those who continue to push to build nuclear reactors right next to the Sundarbans Tiger Reserve, the Nagarjunasagar Tiger Reserve, and the Kanha Tiger Reserve. He would have opposed the World Bank-funded Sardar Saroval Project, part of the infamous Narmada Project that, like China's Three Gorges Dam, eventually plans to displace 1 million humans.

I am not by any means a "Gandhian," because my lifestyle is not by far austere enough. But the more I read his works, the more I become convinced that the "Father of the Indian Nation" was not born to deliver India from the yoke of the British, but rather to deliver Earth itself from foul human ambition.

He would surely have insisted that it should become the purpose of all development to restore health to our ravaged land, restore quality to the water we drink, and productivity to our soils. But this miracle is unlikely to unfold until the consequences that nature delivers forces us to act to survive.

With our water and food security on the verge of collapse, we will ultimately be coerced to turn away from present industrial goals of development. We will be forced to improve generation and transmission capacities of existing power infrastructures, rather than build new projects. We will have to resurface roads, repair culverts, and strengthen shoulders rather than build new highways. We will have to reline canals, improve the condition of the catchment forests of existing dams before building new ones. And we will perforce move to alternate energy options from our drug-like dependence on carbon fuels.

The truth is such options make good long-term economic sense as well, so the sooner we start the long climb back to environmental sanity the better.

Those of us who value and are prepared to defend wildernesses, anywhere in the world, are confronted by crucial and complicated questions that have not thus far been adequately addressed. In which direction does our development destiny lie? How should we balance the needs of people with the imperatives of nature protection? How can we change our heroes so that protectors, not marauders, occupy our pedestals?

Kids for Tigers mobilized 1 million Indian youth, and more help is needed. Photo by Aditya Singh of *Sanctuary*

Trekking through the mountainous Western Ghats forests of Bhimgad in Karnataka, I paused to take in the wilderness vista before me. I was at a height, and thick forests stretched to the horizon all around me. I had just visited the only recorded site in the world of the endangered Wroughton's freetailed bat *Otomops wroughtoni*, and the walk back was hot and strenuous. A rushing crystal pool beckoned and in no time at all the cool waters had washed away dust, sweat, and tiredness. As I bathed, I drank the sweet water and thought to myself how blessed we were. This was the land that Mohandas Karamchand Gandhi had fought to free from the clutches of colonial rule. This was the land that had originally attracted conquerors from afar. This was the land I was born to protect.

Afterword

Cyril F. Kormos

Vice President for Policy, The WILD Foundation

Climate change, the erosion of ecosystem services, an impending extinction crisis, and the steady loss of wildlands around the planet are all spurring conservationists to act at a greater scale, and, if possible, with even greater urgency than ever before. As the conservation community focuses on the enormous challenge of slowing and reversing the trends listed in these pages, wilderness conservation is naturally emerging as an important part of the solution. The recent creation of very large new parks in Canada's Boreal Forest and the Brazilian Amazon, as well as the increased interest in wilderness laws and policies around the world, are some of the best indicators of this trend.

Another good sign is that wilderness conservation is being formally adopted and implemented by a range of sectors. Individuals, indigenous groups, corporations, and combinations of all of the above are now working independently or in partnership with government to protect wilderness values, and to do so explicitly through legal wilderness designations. These designations come in a variety of shapes: from national laws, to municipal ordinances, to tribal ordinances, to easements. But they strive for the same objective—to protect the biological integrity of wild nature for future generations.

As a result of this growing use and acceptance, the term "wilderness" is returning to its fuller meaning. Wilderness areas are not just protected places set aside for solitude, recreation, and the spiritual renewal for urban populations—though these remain important functions. More generally, wilderness areas are wild places that maintain ecosystem services, protect sacred sites, safeguard the homelands of indigenous groups, celebrate peace and cooperation between countries, provide storehouses for biodiversity, create buffers against the effects of climate change, and maintain corridors for traditional migratory peoples, not to mention a range of other essential functions.

These wildlands, free of modern infrastructure and development and largely biologically intact, are not just remote places to be protected for the benefit of the few: they are the living cornerstones of a healthy modern society. As this fundamental realization takes hold around the world, and as we act to protect our remaining wild places, the massive task of slowing the planet's environmental degradation becomes a little bit more manageable.

Appendices

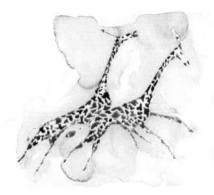

Major Accomplishments,
8th World Wilderness Congress

The 8th World Wilderness Congress was extremely successful, focusing on regional issues in a global context, emphasizing the role of Native peoples in protecting wilderness and wildlands, and also tackling controversial issues such as oil and gas drilling in the Arctic National Wildlife Refuge and global warming. We are pleased to give you a summary of the key accomplishments.

World Wilderness Summit Highlights

The 8th World Wilderness Congress met for nearly two weeks in Anchorage, Alaska, from late September to early October 2005, bringing together 1,200 delegates from up to sixty nations. The Congress achieved all of its conservation objectives and generated several additional, unexpected, and excellent results.

New Wilderness Areas and Legislation Announced:
- Ernesto Enkerlin (President of CONAP, National Commission for Protected Areas in Mexico) announced that "wilderness" will be a new official category of Mexico's protected areas framework.
- CEMEX announced the designation of the El Carmen Wilderness Area on critical biodiversity habitat owned by the corporation in northern Mexico. A management plan was developed with CEMEX partners Agrupación Sierra Madre, Conservation International, BirdLife, The WILD Foundation, and others.
- Jointly Vie Sauvage and the Bonobo Conservation Initiative announced the designation of The Bonobo Peace Forest Initiative in the Democratic Republic of Congo.
- The Mantis Collection, one of South Africa's premiere tourism companies, announced the designation of a new private sector wilderness on their property Sanbona, in the succulent Karoo biome.

Conservation Initiatives:
- The WILD Planet Fund, a funding mechanism for the WILD Planet Project that aims to clearly articulate the economic, biological, and social benefits of intact wilderness
- New inventories and definitions of freshwater and marine wilderness
- The formation of the Native Lands and Wilderness Council and the International League of Conservation Photographers
- The Umzi Wethu Training Academy for Displaced Youth addressing the HIV/AIDS epidemic in South Africa
- Numerous accredited training programs for professionals, scientists, managers, and youth prior to and during the Congress
- Forty-nine resolutions addressing a broad range of conservation concerns

Public Outreach in Alaska:
- Partial and full scholarships for Alaskans attending the Congress

- Free international film festival, "Nature Screen"

- Lasting gift to anchorage of public sculpture by local artist, Rachelle Dowdy

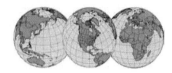

8th World Wilderness Congress Resolutions

Resolution #1

Wilderness and Water: A Basic Human Right

WHEREAS
- Wilderness areas are critical to the survival of life on Earth as the sources of our purest waters;

- The waters that originate from the world's wildernesses are vital to the peoples of the countries through which such waters flow; and,

- The quantity, quality, and seasonality of many water sources in wilderness areas are critically threatened;

THEREFORE, the 8th World Wilderness Congress
- Urges the United Nations to recognize and accept the existence and protection of wilderness areas and the water within and immediately adjacent to them as a basic human right; and,

- Urges that the United Nations works with the governments of the countries of the world to incorporate wilderness protection as a basic human right in their respective national constitutions; and hereby

RESOLVES, to
- Urge the United Nations to recognize the existence and protection of wilderness areas and the water within and immediately adjacent to them across the world as a basic human right;

- Urge governments with de facto wilderness areas, which supply water resources, to protect these areas as de jure wilderness; and,

- Urge national governments to cooperate and act decisively to ensure sufficient water flows within and adjacent to wilderness to sustain basic human rights and wilderness environments across international borders.

Proposer: Bittu Sahgal
Seconder: Mary Ama Kudom, Agyemeng and Farida Shahnaz

Resolution #2
Formation of a Wilderness and Indigenous People Working Group
WHEREAS
- IUCN Guidelines for Protected Area Management Categories, Category 1b, Wilderness Area, defines a wilderness area as a "Large area of unmodified or slightly modified land and/or sea, retaining its natural character or influence without permanent or significant habitation, that is protected and managed so as to preserve its natural condition;"

- It is generally accepted that in many cases indigenous human populations, when present in low densities, have over long periods of time played an integral role in the functioning of their native ecosystems by means of well–adapted livelihoods and cultures; and,

- Vast areas of de facto wildlands exist throughout the world with permanent indigenous human populations living in low densities and retaining much of their traditional livelihood and cultural systems;

- Areas with important biodiversity, ecological and wilderness qualities may be excluded from protection under wilderness designation, solely owing to the permanent presence of human inhabitants; and,

- Those indigenous inhabitants of wild areas, who have been and still seek to be stewards and beneficiaries, as well as dependants upon the ecological health and integrity of de facto wild areas may, as a result of exclusion from de jure protected area status, lack safeguards against exogenous exploitation of de facto wilderness areas;

THEREFORE, the 8th World Wilderness Congress
- Determined that the guiding criteria used to define suitability for designation as "wilderness area" should be re-evaluated with respect to the question of indigenous human communities "living at low density and in balance with the available resources to maintain their lifestyle" on a permanent basis within areas that would otherwise qualify for wilderness designation ensuring always that:

The WWC involves delegates from all sectors, communities, and professions. Here, delegates from a traditional community in Canada and from NGOs in Latin America share their stories in front of the exhibition by the International League of Conservation Photographers. Photo by Carl Johnson

—Their activities do not significantly compromise:

- The health of biodiversity species and habitats

- Ecosystem functioning

- The wilderness character of the area;

—The impacts of their presence are regulated through the mechanism of zonation decided upon through consultative processes with all stakeholders; and,

—That at the outset of any wilderness area planning initiative, limits of acceptable change or other similar models, including population densities, are defined through a consultative process and agreed upon by the indigenous inhabitants of the area; and hereby

RESOLVES, to

- Endorse the formation of an ad hoc working group on Wilderness and Indigenous Peoples, by interested parties acting on a voluntary basis, for the purpose of consultatively reviewing, issues bearing on defining wilderness with respect to indigenous peoples in line with the aforementioned provisos;

- Urge the IUCN Wilderness Task Force, in collaboration with the aforementioned group, to review issues bearing on defining wilderness with respect to indigenous peoples in line with the above provisos; and,

- Request both groups to present their findings both to the 9th World Wilderness Congress.

Proposer: Jonathan HaBarad and Duncan Purchase
Seconder: Evelyn Yobera, Emmanuel Hema, and Shivani Bhalla

Resolution #3
International Definition of Wilderness, Biodiversity Conservation, and Ecosystem Services

WHEREAS

- Recent work has conclusively demonstrated the key role played by wilderness areas for the conservation of biodiversity resources and the provision of ecosystems services as defined by the Millennium Ecosystem Assessment, globally;

- None of the international definitions of the protected area category: Wilderness Area (e.g., of IUCN and the World Wilderness Congress) list biodiversity conservation and the provision of ecosystems services as primary objectives/functions for wilderness areas; and,

- None of the statutes of the various nations that have to date set wilderness systems aside, specify these as objectives/functions for wilderness areas;

THEREFORE, the 8th World Wilderness Congress

- Understands that while the traditional objectives for wilderness (such as the provision of solitude and control of forms of access) are fully recognized, the definition of ecosystems services incorporates these and should therefore be retained. It is essential that vital functions of biodiversity conservation and ecosystems services also be recognized as primary objectives/functions for wilderness areas; and hereby

RESOLVES, to

- Recommend to IUCN that it should amend its definition to include biodiversity conservation and ecosystems services as primary objectives; and also

- Recommend to nations that maintain wilderness systems to amend their legislation accordingly.

Proposer: W. R. Bainbridge
Seconder: Chad Dawson and Malcolm Draper

Resolution #4

Wilderness—A Criterion for World Heritage Listing

WHEREAS

- Under current operational guidelines for the consideration of nominated natural area properties, the conservation of important wilderness values of the area nominated for inscription on the list of World Heritage properties is relegated to being either a factor influencing the biophysical integrity or an element of the conservation of the scenery criterion;

THEREFORE, the 8th World Wilderness Congress

- Endorses that wilderness criterion is necessary under the operational guidelines for the implementation of the World Heritage Convention for natural area properties to reflect the true value and role of wilderness in this increasingly crowded world;

- Identification, recognition, and promotion of wilderness that would follow the establishment of such a World Heritage criterion would provide adequate protection of intrinsic values and more people with life-changing experiences and thereby provide the inspiration to work for a more environmentally sustainable society; and

- A wilderness World Heritage criterion would also help to secure a higher priority for nature-focused management for the reserves listed under that criterion; and hereby

RESOLVES, that

- The 8th World Wilderness Congress request the World Heritage Committee of UNESCO to include wilderness value as one of the criteria in the World Heritage Operational Guidelines for the assessment of natural areas nominated for World Heritage listing.

Proposer: Keith Muir
Seconder: Kevin Kiernan and Haydn Washington

Resolution #5

Protect Arctic National Wildlife Refuge from Oil Development

WHEREAS

- The United States Congress is preparing to vote on a major budget bill that includes an unprecedented authorization to open the coastal plain of the Arctic National Wildlife Refuge to oil and gas exploration and development;

- The Arctic Refuge is the only area protecting a broad spectrum of sub-Arctic and Arctic habitats in the United States, and its coastal plain supports internationally significant resources subject to treaties and agreements, including polar bear denning areas, snow goose autumn staging habitat, nesting, feeding, and molting areas for 135 species of migratory birds, and critical calving and post-calving habitats for the Porcupine caribou herd. This transboundary caribou herd supports U.S. and Canadian indigenous peoples for their subsistence livelihood, and development particularly threatens the nutritional, spiritual, and cultural needs of the Gwich'in people;

- The transboundary protected areas of the Arctic Refuge and adjacent Canadian national parks subject to the Agreement on Conservation of the Porcupine Caribou Herd and the Canadian Government opposes development of the refuge;

- Numerous polls have show that the majority of the American people support continued protection from energy development of this fragile, wildlife-rich area, the only place in America where a complete Arctic ecosystem remains intact from its headwaters to the ocean, for its biological, aesthetic, spiritual, cultural, symbolic, and wilderness values, and for its integral importance to the subsistence way of life and culture of indigenous people; and,

- The U.S., the world's largest consumer of fossil fuels and thus the largest contributor to global climate change, has taken no measures to reduce dependence on fossil fuels and failed to acknowledge the seriousness of global climate change by ratifying the Kyoto Protocol;

THEREFORE, the 8th World Wilderness Congress
- Recognizes that the major impact the U.S. has on global climate change makes it imperative for the U.S. to adopt a sound energy policy, based on conservation and promotion of renewable energy sources to reduce dependence on fossil fuels; and,
- Recognizes the worldwide benefits of preserving and sustaining an intact Arctic ecosystem, which cannot be developed without profound losses to its wild qualities and endangering the future of the Gwich'in people; and hereby

RESOLVES, that
- The United States adopt a national energy policy primarily relying on conservation, energy efficiency, and expanding clean renewable sources of energy, while reducing reliance on fossil fuels and nuclear power, while fully enforcing all existing laws aimed at environmental protection, and to ratify and implement the Kyoto Protocol on global climate change; and,
- United States Congress and President prohibit oil exploration and development in the coastal plain "1002" area, the biological heart of the Arctic National Wildlife Refuge, and to designate this area as wilderness under the Wilderness Act.

Proposer: Betsy Goll and Luci Beach
Seconder: Gladys Netro, Anna Davidson, and George Schaller

Resolution #6
Kamchatka–Alaska Project
WHEREAS
- Among the important issues raised at the 8th WWC was conservation in the Russian Far East, specifically in Kamchatka.

A Kamchatka-Alaska roundtable took place on October 2 in Alaska at Campbell Creek, in which delegates from Kamchatka, representatives of Alaskan nature conservation agencies, representatives from other U.S. states and other countries participated. A list of proposals on development of a partnership cooperation between Alaska and Kamchatka protected areas was worked out and supported by all participants;

THEREFORE, the 8th World Wilderness Congress
- Acknowledges that parties to those proposals would like to see them implemented; and hereby

RESOLVES, to
- Urge development and implementation of a Kamchatka-Alaska Partnership for Protected Areas by the UNDP/GEF.

Proposer: Alexey Stukalov
Seconder: Tatyana Mikhailova and Valery Komarov

Resolution #7
The National Landscape Conservation System Has a Critical Role in Protecting the Deserts of North America
WHEREAS
- The deserts of North America have been identified by Conservation International as one of five global high-biodiversity wilderness areas deserving management designed to protect their relatively undisturbed condition;

- The Department of Interior (DOI), Bureau of Land Management (BLM) administers the National Landscape Conservation System (NLCS), which is a system of conservation areas composed of wilderness, wilderness study areas, wild and scenic rivers, national monuments, national conservation areas, and other special designations;

- There are NLCS units within the deserts of North America composed of 113 separate protected units totaling 8.8 million acres including wilderness, wild and scenic rivers, national monuments, and national conservation areas as well as 282 wilderness study areas totaling 7.8 million acres that are under consideration for potential wilderness designation; and,

- The NLCA provides an existing and viable management mechanism for long-term protection of wilderness and other protected areas on public land within a national system;

THEREFORE, the 8th World Wilderness Congress
- Commends the DOI for establishing the NLCS, which has resulted in a positive impact on protection of the deserts of North America, and hereby

RESOLVES, that
- The BLM is encouraged to:

    —Increase support for management of the NLCS;

    —Emphasize long-term protection of existing NLCS units to maintain their biodiversity and other values;

    —Emphasize training of personnel who manage units of the NLCS;

    —Attempt to increase public awareness of NLCS units and the ecological and economical benefits provided by the NLCS; and,

    —Seek opportunities to expand the NLCS in the deserts of North America by adding additional units representing high biodiversity wilderness.

Proposer: Marilyn Riley
Seconder: George Nickas, Peter Landres, Dave Foreman, Doug Scott, and Steve Ulvi

Resolution #8

Expansion of Nahanni National Park Reserve in Canada
WHEREAS

- Nahanni National Park Reserve only protects a small portion of the South Nahanni watershed and the karstlands of the Ram Plateau;

- The karstlands are the most highly developed sub-Arctic karst in the world with canyons, labyrinths, poljes, and underground rivers;

- A proposed mine at Prairie Creek upstream from the South Nahanni River would threaten water quality and build an access road that would fragment the karstlands;

- The Dehcho First Nations, the traditional people of most of the South Nahanni watershed and the Ram Plateau karstlands, desire the entire area of 36,000 square kilometers to be protected in an expanded national park;

- The South Nahanni River watershed offers one of the finest wilderness canoeing experiences on Earth; and,

- The Government of Canada's Action Plan for National Parks calls for the expansion of the Nahanni National Park Reserve;

THEREFORE, the 8th World Wilderness Congress hereby
RESOLVES, that

- The delegates of the 8th World Wilderness Congress express their enthusiastic support for expansion of the Nahanni National Park Reserve to protect the entire South Nahanni watershed and the karst of the Ram Plateau and call on the Government of Canada to expand the park reserve immediately.

Proposer: Grand Chief Herb Norwegian
Seconder: Jim Thorsell and Rob Buffler

Resolution #9

Supporting the Expansion of Waterton Glacier International Peace Park into British Colombia, Canada's Flathead Valley

WHEREAS

- Waterton Lakes National Park (Canada) and Glacier National Park (U.S.) together comprise Waterton-Glacier International Peace Park, the world's first Peace Park and also a UNESCO World Heritage Site;

- The Flathead Valley is one of the most important areas for large carnivores in North America;

- Glacier National Park protects a vital portion of the Flathead Valley, but there is no corresponding protection upstream on the Canadian side of the border;

- The Government of Canada's Action Plan for National Parks 2002 contemplates the expansion of Waterton Lakes National Park into British Colombia's Flathead Valley;

- The Ktunaxa First Nation, the traditional people of the area, have called for a feasibility study into the proposed expansion of Waterton Lakes National Park; and,

- The Regional District of East Kootenay, representing local governments in the area, has called for a feasibility study into the expansion of Waterton Lakes National Park;

THEREFORE, the 8th World Wilderness Congress hereby RESOLVES, that

- The delegates of the 8th World Wilderness Congress express their enthusiastic support for expansion of Waterton Lakes National Park into 45,000 hectares of the Flathead Valley and urge the governments of Canada and British Colombia to proceed forthwith with an expansion feasibility study.

Proposer: Wayne Sawchuk
Seconder: Gary Tabor and Rob Buffler

Resolution #10

Establishment of a European Wildland Network

WHEREAS

- Many good wildland-related initiatives are operating without linkage or sometimes even knowledge of each other;

- In many areas, organizational capacity to promote opportunities and combat threats to wildland is weak;

- The full range of wildland benefits, and how to value these, is not widely appreciated even among conservation organizations;

- Wildland benefits are often not adequately promoted to policy makers, landholders, local communities, or other relevant interests;

- Much of the present conflict between wildland and biodiversity interests can be resolved with wider knowledge of large-scale natural habitat management techniques; and,

- There is a requirement for coordinated development of a wildland strategy to address threats to wildlands and promote opportunities for their protection and large-scale restoration;

THEREFORE, the 8th World Wilderness Congress

- Encourages that a Europe-wide network should be established to promote knowledge sharing, strategy, and general coordination between the various organizations involved in wildland protection, restoration, promotion, and usage; and hereby

RESOLVES, that

- The above position and calls for the establishment of such a network, to undertake the following activities:

    —Share best practice in a.) usage, valuation, and promotion of the full range of economic, social, and environmental benefits of wildland; b.) restoration and management of wildland; c.) organizational capacity building;

—Represent wildland interests to policy makers at EU level;

—Promote development of a proactive wildland strategy;

—Offer an early warning facility to highlight specific threats; and,

—Support development of a consensus on the value of wildlands among interested parties—including conservationists, landholders, farmers, ecotourism interests, and operators of inner urban social programs.

Proposer: Toby Aykroyd and Jo Roberts
Seconder: Zoltan Kun and Reinelt Arthur

Resolution #11

European Wildland Conference in 2007
WHEREAS

- Many good initiatives in Europe currently operate in isolation from each other;

- In many areas, organizational capacity to promote opportunities and combat threats to wildland is weak;

- The full range of wildland benefits, and how to value these, is not widely appreciated even among conservation organizations working in the field;

- There is scope for a wildland strategy to combat threats to wildlands and take fuller advantage of opportunities for their protection and large-scale restoration; and,

- There is valuable opportunity to agree and share best practice in all aspects of wildland management and promotion;

THEREFORE, the 8th World Wilderness Congress

- Endorses that a conference should be convened within the next eighteen months, in Central Europe, to facilitate interchange between wildland specialists and NGOs, EU and government representatives, together with interested parties from the recreational, landholding, local community, and urban social program sectors, and hereby

RESOLVES, that

- This conference be established and supported within the framework of the World Wilderness Congress structure, to involve the following objectives:

  —To produce a comprehensive review of existing wildland areas and initiatives in Europe;

  —To assess and map potential for restoration of large-scale new natural habitat areas;

  —To agree and share best practice in wildland habitat restoration and management, in usage and promotion of wildland benefits and in development of organizational capacity for promoting wildland interests;

  —To affirm key elements for a coordinated wildland strategy; and,

  —To formally launch a wildland support network.

Proposer: Toby Aykroyd and Jo Roberts
Seconder: Kees Bastmeijer

Resolution #12

Wilderness Areas in Asia
WHEREAS

- Wilderness areas in Asia, particularly on the Tibetan Plateau, the Pamir Mountains, and other mountain regions in western China, Tajikistan, Afghanistan, Pakistan, and India, are insufficiently recognized;

- The Tibetan Plateau serves as headwaters of major Asian rivers (Brahmaputra, Salween, Mekong, Yangtze, and Yellow), which support billions of people downstream, and is one of the world's premier biodiversity areas;

- Is a critical area for the future, with the two most populous countries in the world, both of which are developing rapidly; and,

- The 8th World Wilderness Congress has an insufficient coverage of these and other Asia wilderness areas;

THEREFORE, the 8th World Wilderness Congress hereby RESOLVES, that

- The international conservation community and donor agencies encourage national governments of western China, Tajikistan, Afghanistan, Pakistan, and India to create and maintain these wildernesses in the Tibetan Plateau and the Pamir Mountains, and,

- That at the 9th World Wilderness Congress there should be a much better coverage of Asian wilderness (internal recommendation).

Proposer: Sun Shan
Seconder: George Schaller

Resolution #13

International Support for Protected Areas in Myanmar
WHEREAS

- Myanmar, next to Indonesia, has the richest biodiversity in Southeast Asia;

- Myanmar currently has only 2% of its lands in protected areas in contrast to the approximately 10% of other Asian countries; and,

- Myanmar receives very little international funding to establish additional protected areas to protect its remaining biodiversity;

THEREFORE, the 8th World Wilderness Congress

- Recognizes the urgent need for international funding for the establishment of protected areas in Myanmar; and hereby

RESOVLES, that

- The government of Myanmar be encouraged to establish more protected areas, and,

- The 8th World Wilderness Congress encourages the international community to support the establishment of protected areas in Myanmar.

Proposer: Daniel H. Henning
Seconder: Haven B. Cook and Randall Arauz

Resolution #14

Protection of the Zambezi Valley from the "Chirundu Project"
Agricultural Development
WHEREAS

- It has come to the attention of the 8th World Wilderness Congress that 120,000 hectares of agricultural development is scheduled to start on November 1, 2005 in a contiguous riparian wilderness area of the Middle Zambezi, including Mana Pools (a designated national park and World Heritage Site);

THEREFORE, the 8th World Wilderness Congress
- Understands the need for "food security," but believes that it should be done by taking especial cognizance of the Foundation for Sustainable Development, namely the environment and its adequate protection; and hereby

RESOLVES, to
- Request the Zimbabwean government to:

  —Issue an official statement clarifying the current situation; and,

  —Agree to a full, independent Environmental Impact Process, including wide national and international participation, prior to the project's commencement.

Proposer: Paul Dutton
Seconder: Valerie Purvis and Patty Rees

Resolution #15
Cultivating Wilderness Conservation Values within the Mining Sector: The Langer Heinrich Uranium Mine, Namibia
WHEREAS
- The Namibian government has recently granted approval of a new uranium mine located at Langer Heinrich, within the Namib-Naukluft National Park, to be developed by Paladin Resources;

- The Namib-Naukluft National Park has been identified as a globally important bird area by BirdLife International, and lies within the Namib Desert Wilderness Area, which is identified as a globally important wilderness area by Conservation International;

- This development appears likely to impact the integrity of the Namib-Naukluft National Park and its biodiversity values, both immediately and in the long term; and,

- The mining company involved appears to have no track record in mining and rehabilitating ecologically sensitive areas;

THEREFORE, the 8th World Wilderness Congress
- Expresses its concern that such a mining development should proceed within an outstanding and sensitive location within the Namib-Naukluft National Park, and apparently without adequate plans for impact mitigation and site rehabilitation; and hereby

RESOLVES, that
- The World Wilderness Congress urges the Namibian government to:

  —Undertake a comprehensive action planning process in collaboration with Paladin Resources (and its subsidiary in Namibia, Langer Heinrich Uranium), involving relevant and appropriate local, national, and international stakeholders;

  —Use such engagement to ensure that the impacts of the mine development at Langer Heinrich and associated areas within the Namib-Naukluft National Park are avoided wherever possible and are mitigated where appropriate;

  —Ensure that a closure plan addressing full rehabilitation of the mine site and associated infrastructure impacts is developed and approved through a process of public consultation; and,

  —Pledge to maintain the highest standards in impact assessment and public consultation regarding any further developments of this nature within the Namib-Naukluft National Park and other protected areas of international significance.

Proposer: Jonathan R. Stacey
Seconder: Antonio M. Claparols, Javier M. Claparols, Alec Marr, Professor Graham Kerley, Michael Hoffmann, Matthew Norval, Nyika Munodawafa, Sandy Slater, and Bittu Sahgal

Resolution #16

Saving the Olifants River Gorge in the Kruger National Park

WHEREAS

- The governments of South Africa and Portugal agreed in 1972 that parts of the Kruger National Park, South Africa, could be flooded by the construction of the Massingir Dam in Mozambique (a colony of Portugal at the time);

- The Massingir Dam in Mozambique was never completed (floodgates were not installed) due to the breaking out of war in Mozambique after the colonial power was ousted in 1975;

- The Olifants River Gorge in the Kruger National Park will be flooded and filled with sediment from construction after installation of floodgates on the Massingir Dam in Mozambique continues;

- The Olifants River Gorge in the Kruger National Park constitutes a designated wilderness area and unique river ecosystem, containing the largest crocodile population in southern Africa, which will be lost if this happens;

- The Kruger National Park was never consulted in an environmental impact process and was never given the opportunity to comment on the proposed developments in a structured way; and,

- The 8th WWC recognizes that:

  —Water is a crucial element in the reconstruction of the Mozambique Nation after nearly twenty years of civil war.

  —Mozambique has limited opportunities to develop water storage facilities due to its flat topography and location on coastal plains;

THEREFORE, the 8th World Wilderness Congress

- Understands the need for water development, but believes that the present development is only one option and that there was no impact assessment process to consider alternative scenarios that would be more in line with the principles of the new SA National Water Act, SADC Protocols, or international treaties on shared water resources; and hereby

RESOLVES, to

- Call upon the governments of South Africa and Mozambique to renegotiate the agreement so that water for Mozambique is secured from the more than 2,500 dams in South Africa instead of installing the sluice gates on Massingir Dam and flooding the Olifants River Gorge in the Kruger National Park.

Proposer: Ulf Doerner
Seconder: Dr. Andrew Venter

Resolution #17
African Protected Areas Initiative and Trust Fund (APAI/APATF)
WHEREAS

- African governments have shown a commitment to conserving their natural heritage through adoption of the African Convention on Conservation of Nature and Natural Resources of 1968 and appropriately revised in 2003 to accommodate global change:

  —This commitment has been demonstrated by setting aside approximately 1,200 protected areas to conserve biodiversity and African natural heritage;

  —The African governments have renewed their commitment to African environment and development through creation of the New Partnership to African Development (NEPAD);

  —The NEPAD environment action plan has been endorsed by all African heads of state and governments in 2003, has recognized the need to conserve African's biodiversity through creation of the African Protected Areas Initiative (APAI) to be supported by the African Protected Areas Trust Fund (APATF);

—APAI and APATF were also endorsed by the World Parks Congress in Durban in 2003; and,

—African governments are confronted with urgent problems of health, education, agriculture, and others that create competing priorities that affect their support for the proposed APAI and APATF;

THEREFORE, the 8th World Wilderness Congress
- Supports the initiative of African governments to establish APAI and APATF to enhance conservation efforts and commitment, and herby

RESOLVES, to
- Call upon the international community and all those concerned, in particular donor governments, multilateral and bilateral development agencies, and private foundations, to support the efforts of the African governments to establish APAI and APATF.

Proposer: Michael Sweatman
Seconder: Walter Lusigi, Nkhulani Mkhize, and Andrew Muir

Resolution #18

Tropical Protected Areas
WHEREAS
- Conserving tropical forest ecosystems is of vital importance to our common future, and of special importance to wilderness conservation, mitigation of climate change and other environmental services;
- Our governments have set aside protected area networks of several million hectares of legally protected areas, whose value is in the order of several billion dollars;

- There is an urgent need to eliminate poverty and create ecologically sound income generating alternatives for the populations that live in and around protected areas;

- These populations are a vital component of any long-term conservation strategy;

- Our governments incur high costs for social, environmental, and sustainable land use programs in and around protected areas; and,

- The environmental services provided by tropical protected areas can be quantified, measured, and valued;

THEREFORE, the 8th World Wilderness Congress hereby
RESOLVES, that

- Governments and other donors to the Global Environment Facility, other international organizations, and NGOs should increase their funding dramatically so that management costs can be met and environmental services can be reasonably compensated;

- Existing funding mechanisms should be reformulated so that more funds are spent on activities on the ground, with less bureaucracy and greater efficiency; and,

- The process of designing and implementing conservation and sustainable strategies for tropical forests should include greater participation of the views of tropical countries, their governments, indigenous and local peoples, scientists, and environmentalists.

Proposer: Virgilio Viana
Seconder: Carolyn Rodrigues, Carlos Manuel Rodriguez-Echandi, and Robert Tooley

Resolution #19

Marine Wilderness
WHEREAS
- The land and the sea are linked by:

  —The weather that blows across them;

  —The water that evaporates from the sea, falls on the land, and returns to the sea;

  —The birds that fly above both;

  —The migratory fish that inhabit both; and,

  —And the human species that fish in the sea, inhabit coastal regions, and consume the ocean's bounty;

THEREFORE, the 8th World Wilderness Congress
In gratitude for the presentations and discussions on marine wilderness hereby
RESOLVES, that
- At all future World Wilderness Congresses, the concept of wilderness be expressly understood to include the marine environment.

Proposer: David Rockefeller, Jr.
Seconder: Larry Merculieff and William J. Chandler

Resolution #20

The Importance of Oceans and Their Ecosystems to Life on Earth
WHEREAS
- Oceans are fundamental to all life on Earth;

- Human activity is causing swift and sharp decline of marine species resulting in severe ecosystem stress;

- 95% of the oceans remain unexplored and undocumented; moreover, connections and relationships are poorly understood; and,

- Photography can be the most powerful means of affecting human opinion;

THEREFORE, the 8th World Wilderness Congress
- And the International League of Conservation Photographers recognize the urgent need for more directed use of compelling marine images in visual communications to alter human behavior; and,

- And the International League of Conservation Photographers endorse the recommendations of *Defying Ocean's End,* chapter 10 on "Communications," with emphasis on photography as a critical communications tool; and hereby

RESOLVES, that
- The members of the ILCP should intensify their efforts to produce powerful imagery, especially in the areas of undocumented environments and issues, and develop new, proactive ways to depict the vital connections and vulnerability within ocean ecosystems, while illuminating new ocean science.

Proposer: Tui De Roy and David Doubilet
Seconder: Nora L. Deans, Jeff Bloodwell, Leslie Smith, Miriam Stein, and C. Flip Nicklin

Resolution #21
Freshwater Wilderness Initiatives in Australia: Protecting Queensland's Wild Rivers
WHEREAS
- On Thursday, September 29, 2005, the Parliament of the State of Queensland in Australia enacted Australia's first comprehensive and stand alone legislation to identify, protect, and preserve that state's remaining wild rivers, including an initial nineteen

river catchments identified for wild river protection by the Queensland government in February 2004;

THEREFORE, the 8th World Wilderness Congress
- Commends the Queensland government on this historic initiative to protect freshwater wilderness; and hereby

RESOLVES, to
- Encourage the Queensland government to build on this initiative through the prompt nomination and protection of all nineteen rivers identified by that government in February 2004; and,

- Encourage all other Australian state governments to legislate for the protection of all remaining Australian wild rivers.

Proposer: Alec Marr
Seconder: Antonio M. Claparols and Professor Lawrence S. Hamilton

Resolution #22

Restoring the Great Ruaha River (Tanzania)
WHEREAS
- In 2001 the Prime Minister of Tanzania and the Prime Minister of the United Kingdom jointly pledged to restore flows in the Great Ruaha River by 2010;

- Inefficient irrigation continues to be a problem;

- Degradation of Usangu wetland continues through human induced impacts, (e.g., livestock, fishing, fires, subsistence farms, etc.);

- The ecological integrity of the river and wildlife has been irrevocably altered; and,

- The socio-economic degradation of tourism, power generation, and livelihoods continue downstream of Usangu wetlands;

THEREFORE, the 8th World Wilderness Congress
- Is in full support for reversing the critical situation of the flows of the Great Ruaha River; and hereby

RESOLVES, that
- The following actions be endorsed by the World Wilderness Congress:

    —Active measures on the ground, to improve water flows during the dry season should be supported and enhanced in the catchment areas upstream of Ruaha National Park;

    —Improve dry season water flows by upgrading existing systems;

    —Enforcement of laws from the relevant ministries;

    —Conduct environmental education programs to all stake-holders;

    —Set times for crop growing;

    —Introduce rice varieties and alternate commercial crops that use less water;

    —To asses and introduce measures to improve quality of domestic water of tail-end users through rain harvesting techniques, boreholes, wells, recycling, etc;

    —Determine the volume of water available and what is required for an ecological flow and for the maintenance of national hydropower resource downstream of the Usangu Basin;

    —Periodic reports on the situation of water flow inside the park should be made by Ruaha National Park and other bodies to the relevant stakeholders;

    —Protect the source of the Great Ruaha River in the upper catchment area;

    —Review existing water rights; and,

    —Eradicate exotic tree species from upper catchment area.

Proposer: M. G. G. Mtahiko
Seconder: M. Sweatman and S. Stolberger

Resolution #23

Conservation of Transboundary Riparian Ecosystems in the Jordan River

WHEREAS

- The Jordan River and the Dead Sea are of universal natural and cultural significance and contain biologically rich and diverse ecosystems that are key resting stops along annual migration routes for 500 million birds that fly between Europe and Asia;

- Riparian ecosystems and the Dead Sea itself are endangered by a reduction in water flow from a historical 1.3 billion cubic meters per year to less than 100 million cubic meters per year; and,

- This reduction threatens the ecology of the region and its regulating and supporting ecosystem services, resulting in the significant resources of shared groundwater in Jordan, Israel, and Palestine being at risk of pollution from untreated sewage;

THEREFORE, the 8th World Wilderness Congress

- Endorses the restoration of the flow to the Jordan River will restore valuable riparian habitats, promote a healthier ecosystem, and promote sustainable development; and hereby

RESOLVES, to

- Support the efforts of Friends of the Earth–Middle East (FOEME) in calling for the restoration of a minimum of 300 million cubic meters per year to the Jordan River, through a collaborative transboundary process, to restore riparian ecosystems along the river so that all communities on both sides of the river can share in the benefits of the natural resources in a manner in balance with nature, placing the future of the Jordan River on the international conservation agenda;

- Urge UNESCO, UNEP, IUCN, CI, CMS, and other international conservation and environmental organizations to support Friends of the Earth–Middle East's "Crossing the Jordan" project to restore the flow to the Jordan River through transboundary conservation of water resources, in order to provide technical and financial assistance toward implementing the actions detailed in FOEME's "Crossing the Jordan" concept paper, and aid in research and monitoring of the restoration of the river; and,

- Assist FOEME in efforts to establish the entire Dead Sea area as a Man and Biosphere Reserve (MAB) to aid in the sustainable development of the region and the creation of a regional, joint management plan by Israel, Jordan, and Palestine.

Proposer: Haven B. Cook
Seconder: Chad Dawson and John Romanowski

Resolution #24

Recognition of the International League of Conservation Photographers
WHERAS

- The conservation community has not had an internationally recognized body of professional conservation photographers to work with;

- The very best images illicit the strongest human emotional responses crucial in influencing public on conservation issues;

- The most important conservation issues deserve the highest quality visual imagery available to facilitate the greatest chance of success; and,

- The ILCP seeks to join synergistically with established conservation goals and objectives;

THEREFORE, the 8th World Wilderness Congress

- Recognizes the ILCP as an essential contributor to the furthering of the Congress's goals to create new protected areas and to defend areas already protected; and hereby

RESOLVES, for

- The establishment of the International League of Conservation Photographers as a catalyst for communication between mass media and conservation be recognized by the public and all relevant institutions; and,

- The World Wilderness Congress to collaborate with the ILCP and invite it to participate in further Congresses, including plenary sessions.

Proposer: Cristina Mittermeier
Seconder: Marty Maxwell and Cathy Hart

Resolution #25

International Grant Fund for Ambitious Conservation Photography Projects

WHEREAS

- Far too few professional photographers have access to sufficient funding to do long-term work on important conservation issues; and,

- There is a dire need of professional imagery to cover these issues;

THEREFORE, the 8th World Wilderness Congress

- Recognizes that photography and photographers should be able to work and cooperate for conservation around the world, and hereby

RESOLVES, that

- An international grant fund for ambitious conservation photography projects be created which:

—Is open for applications from photographers, producers, NGOs, and businesses. Open for funding by corporations-major NGOs, government bodies, international agencies, development banking; and,

—Could be administered by the ICLP (or a major conservation organization). Grants could be allocated for major conservation issues, like ecotourism promotion, developing nature photography tourism industry on private lands, covering bush meat trade, illegal logging, climate change, etc.

Proposer: Staffan Widstrand
Seconder: Michele Depraz and John Martin

Resolution #26

Support for Conservation Photography Campaign at China's 2008 Olympics

WHEREAS

- China is at the forefront of the trade in wildlife;

- China is hosting the Olympics in 2008;

- There are local and international NGOs in place working to counteract that trade;

- Those NGOs already have media access and are gearing up campaigns aimed at both the public and the government to stem that trade and would welcome professional coverage with still photos and videos to make their campaigns more effective; and,

- The government of China is keen to maintain a good international image;

THERFORE, the 8th World Wilderness Congress

- Recognizes the need for quality professional images in those campaigns, and hereby

RESOLVES, that
- A team of photographers be brought together to shoot and provide images to those NGOs; and,

- $5,000 be raised from participating NGOs as seed money to coordinate further fundraising and assignments toward the goal of significantly reducing wildlife trade into China.

Proposer: Connie Bransilver and Karl Amman
Seconder: Florian Schulz, Sally Lahm, Theo Allofs, Jaime Rojo, and Sun Shan

Resolution #27
Working Relationship Between Scientists and Photographers
WHEREAS
- Effective conservation is founded on sound science and good communications;

- Science does not always successfully reach the general public or adequately inform policy decisions;

- Photography is a powerful communications tool that is not always properly planned or funded in scientific endeavors; and,

- There exists a lack of interaction, understanding, and exchange between scientists and photographers;

THEREFORE, the 8th World Wilderness Congress
- Recognizes the ability of photography to bridge the gap between conservation science and the general public, and herby

RESOLVES, that
- The International League of Conservation Photographers create an effective interface between scientific and photographic communities; promote the mutual benefits of photographers and

scientists working together; and establish good working practices including those related to initial funding and ethics.

Proposer: Christian Ziegler and Piotr Naskrecki
Seconder: Carlton Ward, Jr., Djuna Ivereigh, Leda Huta, Jeremy
Monroe, Catherine Cunningham, and Richard Edwards

Resolution #28
International League of Conservation Photographers (ILCP)
Advisory Committee for Ethics, Including Traditional People
WHEREAS

- People are a vital part of the ecosystem; and,

- Nature conservation often excludes culture;

THEREFORE, the 8th World Wilderness Congress

- Recognizes that photographers need to acknowledge, respect,
empower, and give audience to under-represented voices; and
hereby

RESOLVES, that

- An International League of Conservation photographers advisory committee that includes indigenous people be formed to
develop a code of ethics that ensures the human component of
the ecosystem is enlisted in the pursuit of conservation.

Proposer: Phil Borges
Seconder: Melissa Ryan and Chris Rainier

Resolution #29

Local Knowledge and Cultural Practices

WHEREAS

- Local people have knowledge and experience like their sacred wisdom, taboos, and cultural practices that could be used by conservation agencies to sustainably preserve wilderness areas;

- This knowledge is waning or disappearing around the world but particularly in developing countries due to the lack of documentation;

- In places where such knowledge is documented, it may be done without any consent and informed participation of local people; and,

- Local cultural knowledge and practices may be used by researchers and companies without any legal protection, recognition, or benefit sharing with the owners;

THEREFORE, the 8th World Wilderness Congress

- Recommends the participation of local peoples must be ensured at each level of wilderness management planning;

- The local indigenous knowledge should be documented only with the informed consent of the knowledge source including the ultimate use; and

- Local people knowledge and cultural practices should be given legal protection, full recognition, and accordingly accrue proper royalties; and hereby

RESOLVES, to

- Urge the IUCN to work with governments with a view to achieving the above three recommended actions in each country through or in partnership with local community-based organizations; and

- That the IUCN report progress of the above program to the 9th World Wilderness Congress.

Proposer: Mamoona Wali Muhammad
Seconder: Basireletsi Maphane, Sydney Allicock, Chandiafira Musorowanrati, Xian Xue, and Albert Lotana Lokasola

Resolution #30

Protection of the Traditional Lands of Aboriginal Peoples
WHEREAS

- Aboriginal peoples have a fundamental connection with the wilderness lands they call home;

- Intact ecosystems are fundamental to aboriginal cultures and health;

- Aboriginal peoples desire to protect their homelands and support sustainable development;

- In many areas aboriginal peoples have lived on the land in a manner that has left large areas intact; and

- In some places like the DehCho area of Canada's Northwest Territories aboriginal peoples desire a permanent land use plan that protects not less than 50% of their traditional territory;

THEREFORE, the 8th World Wilderness Congress hereby
RESOLVES, that

- The delegates of the 8th World Wilderness Congress support Aboriginal peoples in their efforts to protect not less than half their traditional territories where they desire to do so.

Proposer: Grand Chief Herb Norwegian
Seconder: Terry Tanner and Francois Paulette

Resolution #31
Identifying and Applying Traditional Principles of Sustainability
WHEREAS

- Lands and water bodies now politically designated as wilderness are the homelands of indigenous people who have inhabited and used these wildlands for many centuries in balance with intact natural systems;

- Indigenous elders have shared and continue to pass intimate knowledge of their homelands to their youth, including knowledge of sophisticated stewardship practices and traditional harvest technologies that incorporated conservation strategies;

- Traditional knowledge is site specific, growing from intimate engagement with the land and waters that sustain a traditional and customary way of life;

- There is a core wisdom contained within traditional stewardship practices that transcends locality;

- The Native American prophesies foretold a time when Earth

The WILD Foundation and the WWC emphasize that cultural activities can inform and inspire good conservation solutions. The 8th WWC sponsored a sculpture contest with a peer jury, the winning entry of which was completed, installed, and donated to Anchorage. Left to right: Mayor Mark Benich, Janine Oros Amon (Anchorage Convention and Business Bureau), Rachelle Dowdy (winning artist), Vance G. Martin (President, The WILD Foundation), and Michael McBride (Katchemak Bay Wilderness Lodge and 8th WWC senior advisor). Photo by The WILD Foundation would be in great pain and the teachings could be shared among all people; and,

- The time is now;

THEREFORE, the 8th World Wilderness Congress

- Recognizes there is inherent wisdom in traditional knowledge concerning how to sustain intact complex systems in management of wildlands and waters;

- Encourages collaborative wilderness planning and management between agencies, organizations, governments, and the tribal groups whose homelands exist within and adjacent to these wildlands and waters; and

- Supports the effort initiated by the WWC, and continued during the 8th World Wilderness Congress, to identify traditional principles of sustainability and explore their relevance to wilderness resource management in Alaska; and hereby

RESOLVES, to

- Urge the IUCN Wilderness Task Force to:

  —Identify traditional principles of sustainability and explore their relevance and application to wilderness conservation and management;

  —Use both analytical and creative thinking to recognize the core wisdom within indigenous wilderness engagement and explore stewardship practices and technological solutions that would serve wilderness conservation both directly and indirectly; and

  —Report progress and results to the 9th World Wilderness Congress.

Proposer: Nancy Ratner and Larry Merculieff
Seconder: Alan Watson and Dixie Belcher

Resolution #32
Protected Areas and the Convention on Biological Diversity

WHEREAS

- Noting that the adoption of the program of work on protected areas by 188 Parties to the Convention on Biological Diversity is evidence of the commitment of the global community towards conserving viable; and,

- Representative areas of natural ecosystems, habitats, and species as a contribution to achieving the 2010 target of a significant reduction to the rate of biodiversity loss and offers a unique opportunity for global coordinated actions for the conservation of the world's wilderness areas;

THEREFORE, the 8th World Wilderness Congress

- Considering that a number of intergovernmental and international non-governmental organizations—including IUCN, The World Conservation Union and its World Commission on Protected Areas, Conservation International, The Nature Conservancy, World Wide Fund for Nature, BirdLife International, and Flora and Fauna International—have expressed their support to the implementation of the program of work on protected areas; and,

- Recognizing the timeliness of the adoption of the program of work on protected areas of the Convention on Biological Diversity and its relevance to the objectives of the Congress as a useful framework; and hereby

RESOLVES, to

- Encourage all governments and relevant organizations to effectively implement the program of work on protected areas and integrate as appropriate its provisions, into their strategies, plans and programs relating to the protection of the world's wilderness; and,

- Encourage all relevant organizations including funding agencies and the private sector to support governments, with financial resources and other types of support, in their implementation, the program of work on protected areas of the Convention on Biological Diversity, in particular with regard to the protection of wilderness areas.

Proposer: Kalemani Jo Mulongoy
Seconder: Nik Lopoukhine

Resolution #33

Standard International Certification of Sustainable Tourism for Wilderness

WEHREAS

- Ecotourism services in and adjacent to wilderness are provided in an inconsistent fashion through the globe. There is a lack of consensus about appropriate ecotourism practices. While a number of certification protocols exist, there is a lack of coordination among them;

- The 6th World Wilderness Congress resolved to ensure that communication of wilderness sustainability be further communicated through the mass media; and,

- The 7th World Wilderness Congress addressed the issue of ecological sustainability in maintaining wilderness and the inclusion of local people into the venture with sufficient economic interest in order to allow for long-term economic gain and that both can be achieved through ecotourism and keeping areas as wilderness rather than consumptive activities;

THEREFORE, the 8th World Wilderness Congress
- Agrees that tourism providers should seek accreditation on a voluntary basis; and

• This accreditation process should be accomplished through partnerships with governments, NGOs, community businesses, local residents, indigenous communities, and sustainable tourism initiatives already underway; and hereby

RESOLVES, to

• Urge the IUCN WCPA Task Force on Tourism and Protected Areas to:

—Investigate the potential for a global tourism accreditation body with international standards and regional application of best practices, such as "Leave No Trace" and authentic cultural interpretation specifically for wilderness;

—Identify a global tourism accreditation body for coordinated globally sustainable wilderness ecotourism certification; and,

—Carry out the aforementioned through partnerships with governments, NGOs, community businesses, local residents, indigenous communities, and sustainable tourism initiatives already underway.

And also RESOLVES, that

• A report on progress and outcomes be prepared for the 9th World Wilderness Congress; and,

• To urge tourism providers to seek voluntary accreditation in line with the above outcomes.

Proposer: Kat Haber
Seconder: Dr. Marcin F. Jedrezejczak, Till Meyer, et al.

Resolution #34

Wilderness Areas and Gateway Communities—A Global Strategy

WHEREAS

- The World Wilderness Congress seeks to define methods by which wilderness conservation and tourism can provide reciprocal benefits;

- Geotourism principles extend the appeal of a comprehensive, sustainable approach beyond core environmental constituencies;

- A significant proportion of tourists travel to experience a variety of destination assets, cultural as well as natural; and,

- Wilderness areas are likely to benefit from extending the conservation ethic to all forms of destination stewardship, especially in gateway regions;

THEREFORE, the 8th World Wilderness Congress

- Commends the environmental community for its ongoing, pioneering work in sustainable tourism practices and agrees that geotourism provides the holistic platform policymakers need for applying those practices across the board, since an authentic sense of place is known to appeal to beneficial tourism market segments; and

- Recognizes that gateway constituencies will more likely protect natural habitat when they see protection in general as a tourism benefit, whether of wilderness, local historic sites, or endemic cultural assets, and that an array of enriching, sustainable tourism experiences offers a win-win situation for travelers, host communities, and the wildlands from which all benefit; and hereby

RESOLVES, to

- Endorse the principles of geotourism as a strategy for gateway regions and adjacent wilderness areas and wildlands;

- Urge participants to collaborate to adapt geotourism principles (see annex "Geotourism Principles") to the specific needs of gateway regions; and,

- Encourage participants to join in an alliance to help implement the geotourism approach worldwide.

Proposer: James Dion
Seconder: Angela West and Nils Faarlund

Resolution #35

Wilderness Seminar for Government Agencies

WHEREAS

- The Wilderness Seminar for Government Agencies preceding the 8th World Wilderness Congress successfully brought together government employees from over a dozen countries and tribes to discuss their common wildland challenges;

THEREFORE, the 8th World Wilderness Congress

- Urges that government employees continue to share information, strategies, and best practices to support each other in their common objective of protecting Earth's diminishing wildlands; and hereby

RESOLVES, to

- Urge the U.S. Wilderness Policy Council to provide leadership to work closely with the wilderness agencies of the host countries of subsequent Congresses to further improve international coordination and cooperation on wildland protection through a Seminar for Government Agencies similar in scope and general objectives to the one held September 28 through 29, 2005.

Proposer: Karen Taylor–Goodrich
Seconder: Freek Venter, Kruger National Park

Resolution #36

Global Network for Wilderness Information and Stewardship

WHEREAS

- The World Wilderness Congress provides the only open global forum for the wilderness community; and, these opportunities for learning and networking that happen every three to four years are immense and immeasurable;

- Between each WWC convention, there is an urgent need to develop and strengthen opportunities for sharing lessons about wilderness and sharing information that is of utmost importance for rational decision making;

- Relevant tools, both existing and to be developed, include a website, email listservs, and linked relational databases containing education, research, and specific information about wilderness areas around the world; and

- Such processes as regional and thematic workshops, exchange programs for wilderness managers, and print and media resources;

THEREFORE, the 8th World Wilderness Congress

- Commends the efforts of those who have developed existing tools, including www.wilderness.net, the Conservation Commons (www.conservationcommons.org), the IUCN-World Commission on Protected Areas Wilderness Task Force website (http://wtf.wild.org), and IUCN's Protected Areas Learning Network (PALNET—http://www.parksnet.org), and others; and,

- Recognizes the value of simultaneously supporting development of other forums and mechanisms that address topical or regional issues (e.g., the Northern Forum on Arctic related issues); and hereby

RESOVLES, that

- The Wilderness Task Force of the IUCN, co-chaired by The Wild Foundation, and others as appropriate, request financial and other support from government agencies and other funders to develop an umbrella global network to foster international communication and learning on wilderness stewardship.

Proposer: Karen Taylor Goodrich
Seconder: Nikita Lopoukhine, Charles Besançon, Nancy Roeper, David Parson, and Lisa Eidson

Resolution #37

Network to Respond to International Conservation Crisis
WHEREAS

- The public is unaware that now Yellowstone National Park is threatened by legislation to open it to oil drilling;

- Conservation crises occur regularly;

- There is no network to trigger appropriate global response; and,

- Photographers are an essential component of that response;

THEREFORE, the 8th World Wilderness Congress hereby,
RESOLVED, that

- International crisis network be created within the conservation community to disseminate timely critical information to the International League of Conservation Photographers members, appropriate individuals, world media, international groups, politicians, and decision makers.

Proposer: Robert Glenn Ketchum
Seconder: Myron and Bethe Wright, Robin White, Kevin Schafter, Leo and Dorothy Keeler, and Jan Schipper

Resolution #38

The Citizens' Role in Prevention of and Responses to Man-made Environmental Disasters

WHEREAS

- Protection of wilderness and natural resources is the responsibility of all the citizens of the world;

- Priorities of government and industry are often different from those of local citizens;

- Local citizens have vast knowledge of the waters and land in their area and the highest interest in stewardship of those resources;

- Local citizens have the most to lose in the event of a man-made environmental disaster;

- It is in the best interest of industries to work with the citizens of the areas affected by their projects;

- Citizens should participate in designing strategies to prevent man-made disasters and should also have a strong voice in the response to such disasters; and,

- A need exists for a network of citizen groups around the world to share information and support in the event of a man-made disaster that threatens or damages wilderness areas both terrestrial and in the sea;

The 8th WWC supported and hosted the first Native Lands and Wilderness Council. Left to right: Julie Cajune (Confederated Salish and Kootenai Tribes, Montana); Ilarion Merculieff (Aleut, Bearing Sea Council of Elders); Don Aragon (Wind River Reservation, Wyoming); Terry Tanner (CSKT, Montana); and Grand Chief Herb Norwegian (Deh Cho First Nation, Canada. Photo by Vance G. Martin

THEREFORE, the 8th World Wilderness Congress

- Congratulates all the citizens' organizations around the world that have worked tirelessly to influence government and industry to take reasonable and prudent steps to protect against man-made disasters; and hereby

RESOLVES, that

- Extractive industries should support an independent citizens' advisory group that will observe, verify, advise, and inform for the life of projects they start.

Proposer: John Devens

Seconder: Mike Cooper, Lois Epstein, Stephen Howell, Ann Rothe, Rick Steiner, Naki Stevens, and Jonathan Wills

Resolution #39

International Journal of Wilderness Translated Abstracts

WHEREAS

- The World Wilderness Congress, The WILD Foundation, and the *International Journal of Wilderness* are increasingly international in scope and interest; and,

- While many scientists, managers, and wilderness advocates read English as a primary or secondary language, there are many working hard on behalf of wilderness protection for whom this is not the case;

THEREFORE, the 8th World Wilderness Congress

- Urges the *International Journal of Wilderness* (and other publications related to protected areas) to work towards the removal of language barriers in the dissemination of information; and hereby

RESOLVES, to
- Urge the *International Journal of Wilderness* to prepare and include abstracts for all major articles in Spanish and French, in addition to English.

Proposer: William T. Borrie
Seconder: Teresa Cristina Magro and Rick Potts

Resolution #40

Improved Support for Rangers and Conservation Personnel
WHEREAS
- Rangers and conservation personnel, including aboriginal and communal rangers, are essential to maintain the integrity of wilderness areas, not only performing law enforcement duties, but also educating visitors and community members;

- Natural resources are getting scarcer outside protected areas and law offenders are becoming more professional, well-equipped, and more aggressive; and,

- The work of the ranger and conservation personnel is becoming more risky on all continents;

THEREFORE, the 8th World Wilderness Congress
- Recognizes that too often rangers and conservation personnel lack the proper official recognition, lack the required training and equipment commensurate with the level of risks that might be encountered; and hereby

RESOLVES, to
- Urge governments, NGOs, and the private sector involved in protected areas management and the conservation community to recognize the extreme importance of rangers as guardians of the wilderness and to provide them with the best training,

equipment, legal support, and backup they need to perform their duties.

Proposer: Juan Carlos Gambarotta
Seconder: J. Paul Corne, Sonja Kruger, and John Crowson

Resolution #41

Deep Ecology
WHEREAS

- Many approaches to wilderness establishment and protection could be made more effective, including those that bring out spiritual value and deeper approaches and perspectives, including intricate intangibles;

- If more innovative approaches were incorporated, including those that bring in spiritual value, and deeper approaches and perspectives, including intricate intangibles; and,

- Deep ecology involves spiritual environmental approaches that get at deeper questions, values, and issues for spiritual perspectives and solutions;

THEREFORE, the 8th World Wilderness Congress

- Agrees to give serious consideration to incorporating deep ecology and its approaches, education, and training in the World Wilderness Congress programs and activities; and hereby

RESOLVES, that

- Professional and appropriate literature and teachings of deep ecology be made an integral part of the activities and programs of the World Wilderness Congress.

Proposer: Daniel H. Henning
Seconder: Nils Faarlund and Charles Edwandsea

Resolution #42

International Awards System to Honor Wilderness Heroes and Champion Stewards

WHEREAS

- Individuals, institutions, politicians, philanthropists, and communities can and do make outstanding contributions to pioneering and progressing world practice in wilderness protection and sound stewardship;

- International awards can assist in raising the profile and influence of those who are leading the way, and can also provide meaningful reward and feedback to these pioneers and their colleagues/communities; and,

- Success begets success;

THEREFORE, the 8th World Wilderness Congress

- Urges international bodies and benefactors to join together in establishing an international awards system (and possibly linked publication and/or website) to honor and raise international awareness of those responsible for outstanding initiatives and success stories in the protection and sound stewardship of wilderness; and hereby

RESOLVES, to

- Endorse the concept of this initiative and recommends that the proposers of this resolution:

  —Investigate the establishment of an international awards system to honor and reward wilderness heroes and champion stewards;

  —Consider strategies to raise global awareness of the aforementioned leading practices and success stories (e.g., through a publication series and/or website); and,

  —Discuss the potential launch of this initiative at the 9th World Wilderness Congress.

Proposer: Glenys Jones
Seconder: Peter Landres and Danielle Jerry

Resolution #43

Proclamation of Pondoland National Park (Wild Coast, South Africa)

WHEREAS

- The Wild Coast is internationally recognized as one of the world's hotspots of plant endemism;

- The area comprises pristine landscapes, resting habitats for endangered and rare species (e.g., Cape Vultures), archeological sites, et al.; and,

- The initial Environmental Impact Assessment (EIA) was flawed and turned down by the South Africa government;

THEREFORE, the 8th World Wilderness Congress

- Congratulates the South Africa government on its recent announcement of their concept to create a Pondoland National Park; and hereby

RESOLVES, to

- Conduct an independent EIA with appropriate scope for addressing the concerns of a wider spectrum of interested and affected parties (IAP);

- Thoroughly investigate the codependency of the current mining and toll road proposals in the EIA process; and,

- Avoid replication of the ribbon/development as witnessed all along European coast lines (e.g., Spain, France, Italy).

Proposer: Ulf Doerner
Seconder: Sven Kreher and Neil Birnie

Resolution #44

*Twin Sister Partnership of Conservation Areas—Expanding the
Notion of Transboundary Parks*

WHEREAS

- The pressures on wilderness and wildlands (conservation areas) are ever increasing, both locally and globally;

- The tasks to be addressed and managed by respective conserving agencies are of comparable nature; and,

- The concepts of "twin (sister) cities" has proved to be effective and beneficial with regard to knowledge transfer, awareness, and capacity building, only to name a few aspects;

THEREFORE, the 8th World Wilderness Congress

- Agrees that the concept of "twin sister conservation areas" would establish another form of cross cultural, conservational, and educational relationships; and

- Best conservation practices should be shared for comparative biomes; and hereby

RESOLVES, that

- All conservation agencies investigate the potential for twin partnerships of conservation areas; and,

- To establish twin partnerships as an outreach to both public academia government, civil society, and business.

Proposer: Ulf Doerner
Seconder: Walter Lusigi and Douglas Tompkins

Resolution #45

Cultural Outreach Innovation "Ballet and Wilderness"

WHEREAS

- The idea of "Ballet and Wilderness" was conceptualized on the day of the launch of the Wilderness Foundation Germany in Munich on June 26, 2003; and,

- "Ballet and Wilderness" is recognized as an innovative initiative in the field of arts, reaching out to people in a new way and stimulating awareness for the intrinsic values of wilderness;

THEREFORE, the 8th World Wilderness Congress

- Congratulated the Bavarian State Ballet on its innovative initiative, which was first staged outdoors and presented to the public in the forest of the Bavarian National Park in July 2004; and hereby

RESOLVES, that

- The Bavarian State Ballet pursue its initiative called "Ballet and Wilderness" and become a role model for other ballet institutions, ensembles, and choreographers to follow suit both nationally and internationally; and,

- The Bavarian State Ballet be internationally recognized as an ambassadors and opinion leaders for wilderness and The WILD Network (i.e., Wilderness Foundations in the U.S., U.K., R.S.A., and Germany).

Proposer: Ulf Doerner
Seconder: Till Meyer and Kuki Gallmann

Resolution #46

Former Training Grounds of U.S. Army in Germany to Be Designated for Nature Conservation, Preferably as Wilderness Areas (IUCN Protected Area Category 1b)

WHEREAS
- The conversion of training ground of the former Soviet Union into protected areas has been good practice ever since the fall of the iron curtain;

THEREFORE, the 8th World Wilderness Congress
- Agrees that efforts should be taken to identify former U.S. military areas that could qualify for conservation purposes; and hereby

RESOLVES, to
- Encourage U.S. and German officials to jointly identify land use options in particular for conservation; and

- That U.S. land holdings in Germany where military use is expected to discontinue in any foreseeable future should be examined for conversion into protected areas with emphasis on wilderness values.

Proposer: Till Meyer
Seconder: Ulf Doerner

Resolution #47

The World Wilderness Congress and Religion-Based Conservation Groups
WHEREAS
- Much of humanity is losing its connection to the natural world; and,

- When people lose touch with what sustains them, they act in ways that contribute to their own demise by desecrating Earth's life-sustaining ecosystems;

THEREFORE, the 8th World Wilderness Congress
- Recognizes that when people understand and value their relationship to their environment, they tend to behave in ways beneficial to themselves and the planet; and,

- Understands that many leaders in religions and beliefs around the world are forming groups and urging their congregations and followers to protect and care for the creation; and hereby

RESOLVES, to

- Collaborate with religion-based groups from any culture that seek to further environmental goals. The goal of this collaboration is to advance conservation of wilderness by providing information, photography, and publications to these groups.

Proposer: Gary Braasch
Seconder: Amy Gulick and Steve Maka

Resolution #48

Wilderness, Youth, and Young Professionals
WHEREAS

- Today, wilderness is more important than it has ever been, but it is not as important as it will be tomorrow; and,

- The protection and management of wilderness tomorrow will be in the hands of the youth and young professionals of today;

THREFORE, the 8th World Wilderness Congress

- Recognizes that it is an imperative for the generation of wilderness advocates and managers of today to encourage, support, and actively assist youth and young professionals to step into their shoes; and

- This should include aspiring to create a platform from which youth and young professionals can themselves help to make this happen; and hereby

RESOLVES, that

- Future World Wilderness Congresses should have youth and young professionals attending as delegates from each world region (North America, South America, Africa, Asia, Europe, Australia-Pacific);

- The World Wilderness Congress seeks to ensure that those regions are represented by multiple youth and young professional delegates; and,

- Where it is financially or practically difficult for those regions to be well and diversely represented, that financial or practical assistance be provided wherever possible by contributors to the World Wilderness Congress:

  1. That, utilizing these delegates, a permanent role is formed within the World Wilderness Congress for youth and young professionals, including:

     A. Providing a concurrent youth program featuring mentors;

     B. Three or more youth and young professional delegates to each World Wilderness Congress are consulted on the broad program in the early stages of planning;

     C. A reception or other forum is provided at each World Wilderness Congress for youth and young professionals to meet and begin to build relationships with other wilderness advocates and managers.

  2. That the 8th World Wilderness Congress propose that The Wild Foundation provide assistance for a new network of

The WWC encourages youth involvement in international conservation issues. Several young leaders report to the 8th WWC plenary on their work, including Drew Cason, Juhi Chaudhary, Matimba Baloyi, Maan Barua, and others. Photo by Carl Johnson

youth and young professional wilderness advocates and managers by:

A. Providing a website and web forum for youth and young professional wilderness advocates and managers;

B. Having one plenary presentation at each WWC who is a youth or young professional speaking on youth, young professionals, and wilderness.

Proposer: Karen Martinsen
Seconder: Tim Greyhavens and Andrew Wong

Resolution #49

World Wilderness Environmental Television Awards
WHEREAS

- Media coverage often determines whether an environmental effort succeeds or fails;

- Professionals and lay people working for great conservation causes are overworked and under recognized;

- Every crusader for change is inspired and renewed by seeing others overcome their challenges;

- Network television, in combination with Internet and website coverage, will instantly promote causes, not only of the award winners, but those nominated;

- Television and movie celebrities are gaining notoriety for their support of wilderness and conservation causes;

- Celebrities, actors, politicians, and news people worldwide who are nominated for awards will receive a personal or professional benefit from becoming involved in and attending this awards program, thus assuring a strong number of television viewers;

- Awards given to "Kids That Care," whom are voted on by the television audience, will present reality show style participation that will inspire the next generation of conservations and environmental stewards;

- Inspiration generated by the stories and recognition of other lay people will encourage members of the audience to tackle their own local issues, document their results, and enter new reality show categories in future awards programs;

- Awards can be crafted with two goals in mind: 1) to showcase the most important environmental challenges, and; 2) to promote the widest possible television audience, thus insuring the environmental message is heard by the most number of people;

- Major potential sponsors, such as British Petroleum, General Electric, and hybrid car companies, currently advertising their environmental concern will have their target market watching the show and their special advertisements during commercial breaks;

- The organizations that created the Oscars, Emmys, Golden Globes, and the Miss America Contest have demonstrated that license fees derived from their ceremonies can finance a major portion, if not all, of their group's activities; and,

- Television producer Larry Brody, who has written, produced, or created *Star Trek, Voyager, Walker Texas Ranger, Micky Spillane's Mike Hammer, The Fall Guy, Police Story,* and *The Streets of San Francisco,* has expressed great enthusiasm for this project and is interested in offering his production company's services to create, film, assist with licensing the program, and promote the event;

THEREFORE, the 8th World Wilderness Congress
- Recognizes the benefits of creating such an awards program will extend far beyond those activists who are nominated, and for decades;

- Recognizes that the templates exist for duplicating the Academy Awards and reality show format; and

- Current members in the World Wilderness community have the experience, contacts, and wherewithal to make this happen, and that the environmental interest will only grow in time, thus increasing the pools of viewers and the value of the license to air the program, and hereby;

RESOVLES, to

- Recommend that the proposers develop this initiative and assemble a team with the background, experience, financial and signing authority to work out the details, secure the funding, and bring this project to fruition on or before the next World Wilderness Congress, using all available contacts and ingenuity to create the "First Annual World Wilderness Environmental Awards Program," and license it to a network television channel.

Proposer: Dorothy and Leo Keeler
Seconder: Deborah L. Williams and Karen R. Hollingsworth

Murphy Marobe, chairman of the 7th WWC (South Africa) and head of communications for SA President Thabo Mbeki, begins the 8th WWC proceedings by handing over the charge to the co-chairs. Photo by Carl Johnson

Author
Contact Information

Don Aragon
Executive Director
Wind River Environmental
daragon@wyoming.com

Elena A. Armand
United Nations Development
    Program
Russian Country Office
Elena.armand@undp.org

Xanthippe Augerot
Director of Conservation Programs
Wild Salmon Center
xaugerot@easystreet.com

Bradley W. Barr
Senior Policy Advisor
NOAA National Marine Sanctuary
    Program
Brad.barr@noaa.gov

Luci Beach
Executive Director
Gwich'in Steering Committee
Gwichin1@alsaska.net

Ann Melina Bell
Outdoor Recreation Planner
U.S. Fish and Wildlife Service
Ann_bell@fws.gov

Leon Bennun
Director of Science
BirdLife International
leon.bennun@birdlife.org

J. M. Bowker
Research Social Scientist
Forestry Sciences Laboratory, USDA
    Forest Service
mbowker@fs.fed.us

Tom Butler
President
Woodshed Communications
tbutler@gmavt.net

Julie Cajune
Indian Education Director
Ronan Public Schools
Julie.cajune@ronank12.edu

F. Stuart Chapin, III
Institute of Arctic Biology
Terry.chapin@uaf.edu

Tarun Chhabra
Founder
Edhkwehlynawd Botanical Refuge
kwattein@sancharnet.in

H. Ken Cordell
Project Leader/Senior Scientist
Forestry Sciences Laboratory, USDA
    Forest Service
kcordell@fs.fed.us

**William Carey Crane, Chandler
    Van Vorrhis, and Jerry "Dutch"
    Van Vorrhis**
Principals
C2 Invest
Debbie@c2invest.net

Armondo J. Garcia
Vice President for Development
CEMEX
Don.roe@cemexusa.com

Adrian Gardiner
Chief Executive Officer
The Mantis Collection
info@mantiscollection.com

Patricio Robles Gil
President
Agrupación Sierra Madres
patricio@terra.com.mx

Len Good
Chief Executive Officer and
    Chairman
Global Environmental Facility
lgood@thegef.org

Jay Griffiths
Writer
Jaygriffiths@onetel.com

**Gabriel Hoeffer and Fernando
    Morales**
Seri Tribe, Sonora Mexico

The Honorable Walter J. Hickel
Governor
Anchorage, Alaska
wjhickel@gci.net

Michael Hoffmann
Program Officer, CABS Biodiversity
    Assessment Unit
Conservation International
M.hoffmann@conservation.org

Albert Lokosola
President
Vie Sauvage, Kokolopori DRC

Fran P. Mainella
Director
U.S. National Park Service
Jennifer_mummart@nps.gov

Jeffery A. McNeely
Chief Scientist
World Conservation Union
Jam@iucn.org

William H. Meadows
President
The Wilderness Society
Bill_meadows@tws.org

Russell Mittermeier
President
Conservation International
R.mittermeier@conservation.org

Dr. Laura Monti
Assistant Research Professor
Northern Arizona University
Laurie.monti@nau.edu

M. G. G. Mtahiko
Chief Park Warden
Ruaha National Park, Tanzania

Andrew Muir
Executive Director
The Wilderness Foundation South
    Africa
Andrew@sa.wild.org

**Kalemani Jo Mulongy, Sarat Babu
    Gidda, and Marjo Vierros**
Secretariat
Convention on Biological Diversity
Jo.mulongoy@biodiv.org

Grand Chief Herb Norwegian
Grand Chief
Deh Cho First Nations
Sara_mcleod@dechofirstnations.com

Trista Patterson
Ecological Economist
PNW Research Station, U.S. Forest
    Service
tmpatterson@fs.fed.us

Ian Player
Founder
World Wilderness Congress
icplayer@eastcoast.co.za

David Rockefeller, Jr.
Board of Directors
National Park Foundation
kpatras@rockco.com

Bittu Sahgal
Founder and Editor
Sanctuary Magazine
bittu@sanctuaryasia.com

Stephan Schwartzman
Anthropologist/Program
    Co-Director
Environmental Defense
steves@ed.org

Trevor Sandwith
Project Coordinator
Cape Action Plan for the
    Environment
Trevor@capeaction.org.za

Fanyana Shiburi
Board of Directors
The Wilderness Foundation, SA
Andrew@sa.wild.org

Douglass W. Scott
Policy Director
Campaign for America's Wilderness
dscott@leaveitwild.org

Cemci Sose (James Suse)
Councillor
Masakenari Village Council

Michael Smith
Scientist, Freshwater Biodiversity
Conservation International
msmith@conservation.org

Terry Tanner
Project Coordinator
Confederated Salish and Kootenai
    Tribes
terryt@cskt.org

Doug Tompkins
Founder
The Conservation Land Trust
Conservatión Patagonia
pumalin@telsur.cl

Franco Zunino
Founder
Associazione Italiana per la
    Wilderness
Wilderness.italia@libero.it

Barbara Zimmerman
Director, Kayapo Initiative
Conservation International
b.zimmerman@conservation.org

The Zambezi Society
Tchuma Tchato Community
projects@zamsoc.org

Delegates present their initiatives and create collaborative networks at each WWC. Here, in Alaska at the 8th WWC, Dr. Susan Canney (Oxford, UK) and Emmanuel Hema (Burkino Faso) confer on their work to save the desert elephants of Mali, West Africa. Photo by Carl Johnson

8th World Wilderness Congress Organization

Founder and Patron

Dr. Ian Player

Honorary Co-Chairs

We remember with admiration, respect, and fondness the late Jay Hammond (1922–2005), former Governor of Alaska and Honorary Co-Chair of the 8WWC

Walter Hickel, Governor of Alaska (ret)

Byron Mallot, President, First Alaskans Institute

Susan Ruddy, President, Providence Foundation

Clem Tillion, President, Alaska State Senate (ret)

Senior Advisors

Mike Andrews, Chief Conservation Officer, The Nature Conservancy

Dr. Sylvia Earle, Executive Director, Marine Programs, Conservation International

Julie Kitka, President, Alaska Federation of Natives

Dr. Walter Lusigi, Senior Advisor, Global Environmental Facility

Michael McBride, Kachemak Bay Wilderness Lodge

William Meadows, President, The Wilderness Society

Larry Merculieff, Coordinator, Bering Sea Council of Elders

Dr. Russell Mittermeier, President, Conservation International

Dr. Jane Pratt, President, The Mountain Institute (ret)

David Sheppard, Head, IUCN Program on Protected Areas

Mead Treadwell, Senior Fellow, Institute of the North

Bittu Sahgal, Editor, *Sanctuary Asia*

Xan Augerot, Director of Conservation Programs, Wild Salmon Center

Executive Committee

Vance G. Martin, Chair; President, The WILD Foundation

Robert Baron, President, Fulcrum Publishing

Ulf Doerner, President, Wilderness Foundation (Germany)

Cyril F. Kormos, Director, The Wild Planet Project; Vice President for Policy, The WILD Foundation

Harvey Locke, Strategic Advisor, Yukon to Yellowstone Conservation Initiative; Senior Advisor, Canadian Parks and Wilderness Society

Andrew Muir, Director, Wilderness Foundation (South Africa)

Dr. David Parsons, Director, Aldo Leopold Wilderness Research Institute

Dr. Alan Watson, Director, WWC Science Program; Senior Scientist, Aldo Leopold Wilderness Research Institute

Ex officio: Julie Dunn, Director, Financial Development, The WILD Foundation

Resolutions Committee

Susan Ruddy (Chair); Vivek Menon (India); Pamela Miller (Alaska)

8th WWC Secretariat

The WILD Foundation

P. O. Box 20527

Boulder, CO 80308

(3025 47th Street, Boulder, CO 80308)

Tel ++ (303) 442 8811 • Fax (303) 442 8877

www.wild.org • info@wild.org

IT Services, Website Development and Management

(remote and onsite)

Dana Guppy, First Mile Services, Dana@firstmileservices.com.

Dan Toomey, Broadband donated by ATT/Alascom

Media and Press Relations

Brad Phillips, Phillips Media Relations, Brad@PhilipsMedia Releations.com

Kathy Day, Kathy Day Public Relations (KD/PR), Kathy@gci.net

Design and Images

The memorable image of Denali in autumn, chosen as the lead image for all 8WWC material was donated by photographer Cathy Hart ©

8WWC logo designed and donated by Margot Muir

Design and publishing donated by Fulcrum Publishing and Patty Maher, www.fulcrum-books.com

Watercolor images courtesy of Rob McIver and *Heron Dance*, www.herondance.org

8WWC International Film Festival provided by Nature Screen International, M. A. Partha Sarathy, info@naturescreen.org

World Wilderness Congress founder, Dr. Ian
Player. Photo by Carl Johnson

Acknowledgments

Lead Sponsor

The Thoresen Foundation

Patrons

Canadian International Development Agency (CIDA)

Conservation International

Fulcrum Publishing, Inc

National Park Foundation

New York Community Trust

The Christensen Fund

The Ohrstrom Foundation

Trust for Mutual Understanding

U.S. Fish and Wildlife Service

USDA Forest Service

Benefactors
Aldo Leopold Wilderness Research Institute
CEMEX
Ford Foundation
National Fish and Wildlife Foundation
National Forest Foundation
Parks Canada

Contributors
Canadian Boreal Initiative
The Bryant-Crane Foundation
Wilburforce Foundation
The Wildlands Endowment Fund
U.S. Bureau of Land Management
USDA Forest Service International Programs

Supporters
National Parks Service
The Wilderness Society

Additional sponsors of specific workshops, sessions, or products
(In-Kind and Financial)
Many hundreds of people and organizations generously donated
time, services, and targeted funds for specific sessions and products. The following is a vastly incomplete list, with sincere
apologies for the many people and organizations whose names
inadvertently do not appear here.

For substantive content and program involvement:

Mike Andrews	Susan Canney
David Banks	Vietaly Churkin
Mark Begich	Patricia Cochran
Henri Bisson	Natilia Danilina
Josh Brann	Teddy Deane

For substantive content and program involvement, continued

Ken Donajkowski

Iain Douglas-Hamilton

Ernesto Enkerlin

Francisco Filho

Gustavo Fonsesca

Dave Foreman

Kukie Gallmann

Alberta Pereira Góes

Lisa Graumlich

John Haines

Lee Hannah

Emmanuel Hema

Eleanor Huffines

Orville Huntington

Glenn Patrick Juday

Graham Kerley

Janus Kober

Adam Linn

Todd Logan

Nikita Lopoukhine

Kathy MacKinnon

Alec Marr

Tambi Matimbo

Mark McDonald

Murphy Morobe

Richard Nelson

Jim Nollman

David Parsons

Rosimeriry Portela

Glenn Prickett

David Quammen

Chris Rainer

Ray Rasker

Willem van Riet

Libby Roderick

Carlos Manuel Rodríguez

Caroline Rodríguez

Helvi Sandvik

Joel Sartore

Peter Seligmann

Windsor Shuenyane

John Shultis

Joe Singh

Mirran Smith

Jonathan Solomon

Simpiwe Somdyala

Vsevolod Stepanitsky

Alberta Stephan

Doug Stewart

Mead Treadwell

Willem Udenhout

James Valentine

Vergilio Viana

Sheila Watt–Cloutier

Kathy Wilkenson

Deborah Williams

Margaret Williams

Krystyna Wolniakowski

Andrew Wong

For substantive content and program involvement, continued

Agrupación Sierra Madre
Alaska Youth for
　Environment and Action
Aldo Leopold Wilderness
　Research Institute
Arkive
Alaska Stock
Alaska EdVentures
Alaska Pacific University
Alaska Moving Image
　Preservation Association
ATT/ALASCOM
Bateleur's—Flying for
　Conservation in Africa
Ben & Jerry's
Blue Earth Alliance
Blue Pixel
Circumpolar Conservation
　Union
Fine Print Images
First Alaskans Institute
Ford Foundation
Fry Communications
Global Environmental
　Facility

International League of
　Conservation Photographers
International Tropical Timber
　Organization (ITTO)
Lufthansa
Marine Photo Bank
National Geographic Society
Nature's Best
Nature Screen
NE Wilderness Trust
Outdoor Photographer
Project Lighthawk
Rockefeller Philanthropic Group
State of the Salmon Consortium
Sweetwater Trust
The Nature Conservancy (Alaska)
Wildlands Conservation Trust
Wild Britain Initiative
Wilderness Foundation (UK)
Wilderness Foundation (SA)
Wilderness Foundation
　(Germany)
Wilderness Leadership School
　(South Africa)
Wild Salmon Center
WWF/Canon

Cathy Hart
Anne Hanley
Cathy Day
Dana Guppy
David Rockwell

Dan Toomey
Gigi Coyle and Win Phelps
Herb Norwegian
Janine Amon
John and Marilyn Hendee

For substantive content and program involvement, continued

Julie Cajune	Marti Marshall
Karen Martinsen	Patty Maher
Karen Ross	Peter Stranger
Larry Kopald	Philip and Mikela Tarlow
Louise Aspinall	Terry Tanner

And special thanks to Michael McBride, Michael Thoresen, Magalen Bryant, Bob and Charlotte Baron, Michael, Diane, and Shannon McBride, Cristina Mittermeier, Patricio Robles Gil, Gary Tabor, Partha Sarathy, Janine Amon, Mary Wagner, Karen Taylor Goodrich, Marshal Jones, Jim Kurth, Julie Dunn, Toni Walker and Logistics, Krista Petumenos, Valerie Purvis, Patty Rees, Patty Maher (for all the constant design help), Emily Loose (for her editorial help and positive attitude), and Carol Batrus (for everything!).

Notes

Chapter 4 Notes
Patterson

1 Actual typologies vary, however. See Boyd and Banzhaf, 2005; Costanza, et al., 1997; de Groot, et al., 2002; Alcamo, et al., 2003; Heal, et al., 2005; Brown, et al., 2006 for review

2 Welfare economics and social justice applications of environmental economics offer some exceptions, however.

Chapter 5 Notes
Barr

1 Marine Theme Roundtable Panel: Peter Auster, Univ. Conn. (U.S.); Bill Chandler, MCBI (U.S.); Donita Cotter, USFWS (U.S.); Gary Davis, NPS (U.S.); Jon Day, GBRMPA (AUS); Andrew Gude, FWS (U.S.); Sabine Jessen, CPAWS (CAN); Luciano Parodi, Chilean Embassy; Greg Siekaniec, FWS (U.S.); Norm Sloan, Parks Canada (CAN); Dan Suman, Univ. Miami (U.S.); Lauren Wenzel, Nat'l. MPA Ctr. (U.S.); Brad Barr, NMSP (U.S.), Moderator

Chapter 8 Notes
Zunino

1 Italian Ministry for the Environment and territory. 2005, Five Years to Make a Difference, Italy launches countdown 2010 initiative to halt the loss of biodiversity.

2 Ministry of the Environment, Nature Conservation Service. 1998. Italian National Report on the Implementation of the Convention on Biological Diversity, p.5.

3 Zunino F. 1980. Wilderness, a new necessity for the preservation of natural areas, National Department of Agriculture and Forests, Italy (in Italian).

4 Zunino F. 1984. A Wilderness Concept for Europe in Martin, V. and Inglis, M. eds, Wilderness The Way Ahead: Proceedings of the 3rd World Wilderness Congress, Finhorn Press and Lorian Press. pp. 61–65.

5 For more information on particular areas designated as wilderness in Italy please see Zunino F. 1995. The Wilderness Movement in Italy: A Wilderness Model for Europe, International Journal of Wilderness, Vol. 1, no. 2, 1995.

6 Friuli Venezia Giulia Region, Regional Executive Board (Delibera della Giunta Regionale) Decree no. 3117/2006.

7 Leopold A. 1921. The Wilderness and its Place in Forest Recreation Policy, Journal of Forestry, reprinted in: Flader S., Callicott J.B. 1991. The River of the Mother of God and other Essays by Aldo Leopold, University of Wisconsin Press, 78.

Chapter 10 Notes
Scott

1 Except where noted, all acreage data in this paper are from www.wilderness. net, accessed January 15, 2006. This is the definitive source for statistics about the NWPS and each of the 680 areas protected by the Wilderness Act.

2 Subsection 2(a), Wilderness Act, 16 U. S.C. 1131 (a). The full text of the act can be found at http://leaveitwild.org/psapp/view_art.asp?PEB_ART_ID=55

3 Kenneth A. Reid, "Let Them Alone!" Outdoor America 5, no. 1 (November 1939), page 6.

4 Howard Zahniser, "The People and Wilderness," The Living Wilderness 28, no. 86 (Spring–Summer 1964), page 39.

5 The way in which this requirement of passing a new law to alter wilderness area boundaries or protections serves to strongly guarantee that preservation will be sustained may not apply as strongly in parliamentary systems, where a change of government by the election of a new legislative majority can result in rapid and wholesale revision of earlier statutes.

6 Aldo Leopold, "The Last Stand of the Wilderness," American Forests and Forest Life 31, no. 382 (October 1925), page 603.

7 Robert Marshall, "The Problem of the Wilderness," The Scientific Monthly, (February 1930), page 146.

8 Zahniser Howard. "A Statement on Wilderness Preservation in Reply to a Questionnaire" [submitted to the Legislative Reference Service, Library of Congress], March 1, 1949, reprinted in National Wilderness Preservation Act, Hearings before the Committee on Interior and Insular Affairs, U.S. Senate (85th Congress, 1st session), June 19–20, 1957, page 192.

9 Senator Hubert H. Humphrey, "Wilderness Preservation," Congressional Record, June 7, 1956.

10 Aldo Leopold, "The Last Stand of the Wilderness," American Forest and Forest Life, 31, no. 382 (October 1925), page 603.

11 Statement of Senator Clinton P. Anderson, in Senate Committee on Interior and Insular Affairs, Wilderness Act, Hearings before the Senate Committee on Interior and Insular Affairs, U.S. Senate (87th Congress, 1st session), February 27–28, 1961, page 2.

12 "Supplementary Statement of Howard Zahniser, Executive Director of the Wilderness Society," in National Wilderness Preservation Act, Hearings before the Committee on Interior and Insular Affairs, U.S. Senate (88th Congress, 1st session), February 28 and March 1, 1963, page 68.

13 George Marshall, Introduction, Wilderness and the Quality of Life, Maxine E. McCloskey and James P. Gilligan, editors (San Francisco: Sierra Club, 1969), pages 13–15.

14 Subsection 4(b), Wilderness Act

15 This is explored in detail in Douglas W. Scott, "Untrammeled, Wilderness Character, and the Challenges of Wilderness Preservation," Wild Earth,

Fall/Winter 2001–2002, pages 72–79, http://www.leaveitwild.org/reports/Untrammeled_Article.pdf.

16 Subsection 4(c), Wilderness Act

17 Preservation of Wilderness Areas, Hearing before the Subcommittee on Public Lands, Committee on Interior and Insular Affairs, U.S. Senate (92nd Congress, 2nd session), May 5, 1972, pages 61–62.

18 http://www.wilderness.net/index.cfm?fuse=MRDG

19 My friend and long-time colleague Dave Foreman summarizes the etymology of the word "wilderness" as traced by historians and philosophers: "According to historian Roderick Nash, the word "wilderness" comes from the Old English Wil-deor-ness, which he defined in 1967 as "place of wild beasts." *Wil*: Wild, or willed. *Deor*: Beast, or deer. *Ness*: Place, or quality. In a 1983 talk at the Third World Wilderness Conference in Scotland, philosopher Jay Hansford Vest also sought the meaning of wilderness in Old English and further back in Old Gothonic languages. He showed that wilderness means "'self-willed land' … with an emphasis on its own intrinsic volition." He interpreted der as of the, not as coming from deor. "Hence, in wil-der-ness, there is a 'will-of-the-land'; and in wildeor, there is 'will of the animal.' A wild animal is a 'self-willed animal'—an undomesticated animal—similarly, wildland is 'self-willed land.'" Vest shows that this wilfulness is opposed to the "controlled and ordered environment which is characteristic of the notion of civilization."

20 Howard Zahniser, "Guardians Not Gardeners," Editorial, The Living Wilderness, Spring/Summer 1963, page 1.

David Rockefeller, National Parks Foundation. Photo by Carl Johnson

Project LightHawk flew media and decision makers to overview Alaskan wilderness issues. Photo by Kurt Johnson

Some of the participants from 26 nations in the accredited Wilderness Management Training Course. Photo by The WILD Foundation

Tlingit drummers open the 8th WWC proceedings. Photo by Carl Johnson

Sheila Watt-Cloutier, Inuit Circumpolar Conference, Canada. Photo by Carl Johnson

Nikita Lopoukhine, Chairman, World Commission on Protected Areas, IUCN. Photo by Carl Johnson

Fran Minella, Director, US National Parks Service. Photo by Carl Johnson

Michael McBride (senior advisor, 8th WWC) and wife, Diane McBride, of Katchemak Bay Wilderness Lodge prepare an Alaskan meal for one of the numerous post-congress training and excursion trips. Photo by Vance G. Martin

Siging an accord for wilderness conservation and collaboration in Mexico. Left to right: Armando Garcia (Executive Vice President for Development, CEMEX); Patricio Robles Gil (Sierra Madre); Russell Mittermeier (Conservation International); Vance G. Martin (The WILD Foundation); Leon Bennun (BirdLife International); Ernesto Enkerlin (CONANP, Mexico). Photo by Carl Johnson

Photographer James Balog (left) and others in front of the first exhibition of the International League of Conservation Photographers, founded at the 8th WWC. Photo by Carl Johnson

Byron Mallott, 8th WWC Co-Chair, Tlingit Nation, First Alaskans Institute. Photo by Carl Johnson

Marshall Jones, Deputy Director, U.S. Fish and Wildlife Service. Photo by Carl Johnson

The resolutions committee worked day and night to negotiate and finalize the 49 resolutions adopted in plenary. Left to right: Vivek Menon (India); Pamela Miller (Alaska, U.S.); Susan Ruddy (Co-Chair of the 8th WWC in charge of the resolutions process); Ilan Lax (South Africa); Frances Gertsch (Canada); Susan Oliver (UK). Photo by Vance G. Martin

Doug Tompkins, Conservación Patagonia, Chile. Photo by Carl Johnson

Youth leaders Matimba Baloyi (South Africa) and Juhi Chaudhry (India) hard at it in the delegate's work room. Photo by Vance G. Martin

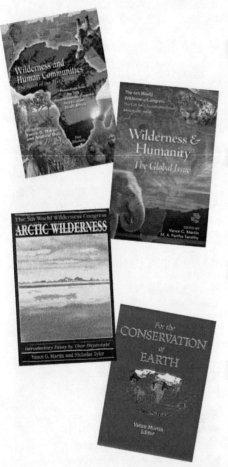